A FRAYED
NEW WORLD

FROM SCIENCE FICTION TO
SOCIETY

DAMINI RANA

Contents

Acknowledgements

I am thankful to Isaac Asimov for inspiring me to write this book and introducing me to the wonderful world of science, and science fiction.

I am also grateful to my parents for always encouraging me to read and supporting my late-night writing sessions with a steady supply of hot chocolate. This book wouldn't be here without you.

CHAPTER 1

INTRODUCTION

I have always loved science fiction, ever since I was allowed to read and watch it. It made me laugh, it made me cry, even scared me at times (literally—*War of the Worlds* was too scary a movie for my 9-year-old self). And it made me *think*.

The term 'science fiction' conjures an image of aliens, laser guns, and an unrealistic dystopia, a medley of wild, improbable scenarios and possibilities. We can like it, we can hate it, but we can't ignore it. We can't ignore the impact it has on our world, from influencing scientific and technological innovations, to shaping public policy. And while it may be easy to dismiss sci-fi as fantasy or imagination, it isn't as easy to dismiss the questions it raises around ethics, or the societal impact of science. As we move into a future fueled by rapid technological advances, these depictions of futuristic (and often dystopian) societies reinforce the need for safeguards and socially acceptable guidelines that innovations in science and technology should follow. They spark public debate on what our future could look like, more importantly what *we want* our future to look like.

As we enter a transitory phase where the science of today catches up with the science fiction of yesteryears, we are experiencing that a *que sera sera* approach to

1

our future will not work. Our response to two critical issues—climate change and the rise of big tech—demonstrates that we haven't yet developed the ability to respond to the challenges we face today, at the speed and scale required. We are running blind, enjoying one invention or application after another, without thinking through the long-term impact of each new innovation. Eventually, when we reach the end of the line, it will be impossible to let go of our addictions to misguided technological developments akin to what we are witnessing with respect to our dependence on fossil fuels. It will be challenging to forge a deep meaningful interlinkage between science and society.

In this book I have discussed how science fiction plays an important role in crafting a symbiotic future between science and humanity, ruminating on the ideas and ideals sci-fi introduces along with the warnings it delivers, specifically highlighting the significance of ethical design.

This introduction, and in some ways a summary of the insights of this book, is structured as follows.

1.1 Overview

Science has transformed life dramatically over the last 150 years, and mostly for the better. Humans are living longer, healthier, and have a better quality of life than ever before. The last two decades have also witnessed an exponential increase in the pace of changes and developments within science and technology. Advancements in computers, communications, biomedical science, and space make it likely that the futuristic worldview of a hyper-connected society, robots, autonomous transport, ultra-cheap energy, and genetically modified humans will become commonplace, if not reality, in the near future.

As machines and artificial intelligence take over not just physical but also most cognitive tasks from humans, there will arise hard questions around the role, economic relevance, and even existence of humans. This transition, and our answers to these questions will irrevocably transform how society functions, be it the moral code that binds individuals within society or the socio-political structures that will emerge to manage these new paradigms.

Moreover, science having had a largely positive effect on humans so far does not necessitate that this trend will continue in the future. It is ostrich-like to assume that the future will pan out smoothly, with solutions popping up at the same pace that problems do, or that there is some prescient think-tank or elected representative or tech entrepreneur who is thinking through these questions, and consciously keeping the collective good of society in mind. We need to arrive at a consensus on what we want the impact of this transition to look like and figure out how we intend to regulate it. And a good place to start this exercise seems to be literature which does exactly that, by presenting variations of our future—Science Fiction.

There is of course, a wide range to sci-fi, a stark contrast between the hopeful and exciting intergalactic worldview of *Star Wars*, the comedic irreverence of *the Hitchhiker's Guide to the Galaxy*, and the sobering worldview of *1984, Brave New World, Fahrenheit 451*, and the like. But regardless of the sub-genre of the book (or the movie), I have always marveled at the authors' ability to envision a futuristic society and create a facile story within its context.

Even if authors liberally employ their creative license to define scientific progress, even if the snippets of the future they present to us are flawed, they still give a cohesive perspective on opportunities and challenges that may emerge. For instance, books like *1984* and *Jurassic Park* have had a huge impact on public consciousness, and shaped socio-political perspectives on censorship and genetic engineering respectively. Science fiction not only shows us what our 'brave new world' could potentially look like, but also cautions us against treading certain avenues, and illustrates obstacles and outcomes we must be prepared for.

Keeping this in mind, I have compiled different perspectives and worldviews from some select works of science fiction, and analyzed their relevance today and takeaways for our present world. Since it would be impossible to review every influential work of science fiction, I have narrowed the list down to nine of my personal favorites:

1. Brave New World by Aldous Huxley, 1932
Dystopian world in 2540 where societal happiness is ensured through planned procreation and drugs leading to a happy but banal existence.

2. I, Robot by Isaac Asimov, 1950
Collection of short stories on the theme of interaction between Robots and Humans, delving into "Robopsychology", and musing on ethics for artificial intelligence.

3. The Moon is a Harsh Mistress by Robert A. Heinlein, 1966
Lunar colony that rises in revolt against Earth. Used to showcase alternate ways to structure politics and society with a fresh start on a lunar colony.

4. Do Androids Dream of Electric Sheep? by Philip K Dick, 1986
Set in Post-apocalyptic San Francisco, follows the story of a bounty hunter on a mission to retire "rogue androids".

5. The Handmaid's Tale by Margaret Atwood, 1985
A dystopian society where the status of women has been relegated to their reproductive function only, set amidst a

strict hierarchical class structure.

6. Jurassic Park by Michael Crichton, 1990
An amusement park with real Dinosaurs recreated using genetic engineering. Demonstrates how things can go wrong when we interfere with natural hierarchies.

7. Snow Crash by Neal Stephenson, 1992
A dystopian, cyberpunk novel set in the late 21stcentury where people spend considerable time as avatars in a metaverse, and how that affects their psyche.

8. The Martian by Andy Weir, 2011
Outlines Mars Missions through the story of one such mission gone awry when one of the astronauts gets left behind and spends over 500 days alone on Mars.

9. Ready Player One by Ernest Cline, 2011
People forego their lives in the real world in favor of virtual identities in a makeshift virtual world, the OASIS, where our protagonist goes on a quest to inherit its creator's fortune.

1.2 Evolution of Science Fiction

The encyclopedia defines science fiction as literary work that deals principally with the impact of actual or imaginary science upon society or individuals.

The earliest references to sci-fi are found in mythology and literature dating as far back as major Indian and

Greek epics. Fantastical stories with flying machines, remote communication, sophisticated gadgets, and time travel were quite common. But these were mostly portrayed as beyond the world of ordinary mortals and reserved for gods, kings, and superhumans.

At the advent of the 16th century, when science fiction was in its early stages, people were fascinated with discovering space like in Sir Thomas More's novel *Utopia* in 1516.

It wasn't till the late 19thcentury that science fiction gained momentum as a genre, as progress within science and technology increased dramatically, and its influence on society began outweighing that of philosophy and religion. In a few decades (interestingly enough, within a single generation) electricity, telegraph, automobiles, airplanes, and skyscrapers transformed the world and dramatically changed the day-to-day existence of people including how they lived, worked and interacted with each other. The best part was that most of these wonderful inventions and conveniences were available to the common man, without socio-economic status significantly affecting access to these technologies. A collective sentiment of humans being able to achieve anything through science, and make any concept a reality if they set their mind to it, gradually started to emerge in this period.

This led to a desire to speculate about possible new elements that advancement in science could bring to society and gave rise to a new genre of literature - science fiction. Some recurring themes included robots, space travel, time travel, genetics, and virtual society. Various works of science fiction provided differing, but interesting perspectives on the consequences of technology on society. From *Frankenstein* in 1817 to *Ready Player One* in 2011, science fiction has morphed with evolutions in science, and has consistently had a tremendous impact on society and scientific innovation. H.G. Wells, Jules Verne, Isaac Asimov, Arthur C. Clarke, Robert Heinlein, Neal Stephenson, and many others are not just considered great authors, but also futurists who have indirectly or directly had an influence on the strategies and policies adopted by corporations and governments.

1.3 Inspiring the World around Us

Visualize a new color. Impossible, isn't it? But if I were to show you one, then it would be easy to do what was seemingly impossible, moments ago. Humans find it difficult to come up with or visualize something entirely new, to conjure ideas out of thin air. Borrowing from Mark Twain,

"There is no such thing as a new idea. It is impossible. We simply take a lot of old ideas and put them into a sort of mental kaleidoscope. We give them a turn and they make new and curious combinations. We keep on turning and making new combinations indefinitely, but they are the same old pieces of colored glass that have been in use through all the ages."

The same principle extends to innovation in science and technology as well.

For example, if we were to wake up tomorrow and be greeted by the headlines '*Invisibility: Now A Reality with Mr. Doe's machine*', we would accredit him with the invention of invisibility. However, invisibility has been a well-known concept, with a plethora of movies, books, and shows based on it. So, did Mr. Doe truly 'invent' invisibility, or did he merely concretize an idea first introduced to us through H.G. Wells' *The Invisible Man*? In a similar fashion, many modern inventions are a physical manifestation of a pre-existing concept or idea. Take smartphones, for instance, their roots being traced to Captain Kirk's communicator in *Star Trek*. 'Futuristic' work taking place in various fields of science is not original in the true sense of the word, but is rather an attempt toward creating ideas we are familiar with, be it SpaceX's attempts towards establishing an interplanetary society (popularized through George Lucas's *Star Wars*), or investment in genetic mutation and engineering (Michael Crichton's *Jurassic Park*: the author does well to warn us that control over nature is an illusion, refer to chapter seven). These are now colloquial concepts which are relatively easy to envision, but only because we are well acquainted with them.

No doubt, many of these ideas and concepts, the 'pieces of colored glass' for our kaleidoscope, are borrowed from or influenced by science fiction. Science fiction and scientific innovation have been deeply interlinked ever since science fiction came up in the 19th century. In fact, it would not be inaccurate to say that the craziest scientific and technological innovations often emerged from these artistic visions. Fiction gives authors a safe space to dream, experiment, and weave a tale without the constraints of reality or laws of physics, and scientists found inspiration in these tales.

An interesting example is the pacemaker, believed to be inspired by Mary Shelley's novel *Frankenstein*. When electricity was invented in the latter part of the 18[th]century, it piqued the interest of various scientists who carried out all sorts of experiments—like twitching a dead frog's legs by passing electricity through it. This process came to be known as 'Galvanism', the application of electric current through biological organisms. Italian Physicist Giovani Aldini (1762-1834) worked extensively with galvanism and created quite a spectacle by conducting experiments on corpses of criminals as public performances. He would animate corpse's by passing current through a corpse's head. The muscles and jaws of the corpse would contort, sometimes even opening an eyelid, and its limbs would violently start shaking. It is believed that this inspired Author Mary Shelley to write the novel *Frankenstein* in 1831. She described the process of galvanism as a key influence in coming up with the core concept of her book. On the basis of this, she wrote about the secret life of Dr. Victor Frankenstein who assembled the body parts of dead corpses and passed an electric current through them to bring them to life, thereby creating his monster (often referred to as Frankenstein, as opposed to Frankenstein's monster), and is one of the earliest works of science fiction documented.

A movie based on Shelley's book was released in 1937, and inspired engineer and inventor Earl Bakken to invent the first battery-powered, wearable cardiac pacemaker in 1957. The device sent out electric impulses to set the rhythm of irregular heartbeats. Earl Bakken worked with both electricity and medicine and founded the company Medtronic.

"What intrigued me the most, as I sat through the movie again and again was the creative spark of Frankenstein's electricity," Bakken said in his 1990 autobiography *One Man's Full Life.*

And that's how a life-saving medical device was inspired by a sci-fi monster.

Not all of us have experienced or seen a pacemaker, so let's take an example of something we can all relate to. Arthur C. Clarke's *2001: A Space Odyssey,* published in 1968, introduced us to one of the most intelligent robots of that time frame—Hal 9000 (abbreviated for Heuristically Programmed Algorithmic Computer). Hal could think, sing, play games, and execute voice-based commands. In today's context, don't these features remind you of your personalized artificial intelligence based smart assistants—Siri and Alexa?

The list of inventions inspired by science fiction is endless, as science fiction introduces and showcases new ideas and possibilities. Long before autonomous cars became a reality, they were popularized by the *Knight Rider* series in 1982. And who knows if it's only a matter of time before we see flying cars become commonplace, akin to the 1997 Hollywood blockbuster *Fifth Element*?

Not all Science Fiction is positive or inspiring or about future inventions, space, and aliens! Margret Atwood's *The Handmaid's Tale*, feminist literature published in 1986 was a bleak outlook on the future based on social changes resulting from the harmful effects of radiation in the environment and chemical poisoning in the food chain. Many argue that this book isn't really science fiction, though I beg to differ. It is very much a utopian vision about the future based on radical social restructuring (more about this in chapter six).

1.4 Sci-Fi is Predictive

"Today's science fiction is tomorrow's science fact" -Isaac Asimov

Many argue that science fiction does not predict the future, science fiction *is* the future.

The future is built on ideas sci-fi introduces, accounts for catastrophes and mishaps sci-fi contains, ruminates on warnings sci-fi delivers, and chases the visuals and ideals sci-fi provides. These ideals are multifold: there are broader themes, such as hope for a better society and what a 'better' society may look like, there are depictions of how technology can dramatically alter societal structures and human psyche, and in addition to these bizarre yet probably concepts, there are tidbits of innovations littered throughout science fiction. Science fiction has inadvertently shaped our present and will influence our future. There will be unimaginable technologies (our eighth color) that crop up as science progresses, but the precursor to these may be found in science fiction.

Sci-fi authors spent time thinking of technologies which did not exist in their lifetime, and while they may not have been exactly on the mark, they weren't off by much either. While we may not have sent men to the moon in a canon like Jules Verne did in his 1865 novel *From Earth to the Moon*, we did eventually send people up there, and the similarities between Verne's hypothetical mission and NASA'S Apollo 11 mission (launched more than a century after Verne's!) are uncanny. Both Verne's and NASA's vessels were made of hollow aluminium shells, both missions were manned by three people, and both landed back into oceans.

Consider this: In 1988, Isaac Asimov predicted that everyone will own personal computers which will be connected to a repository of information and books. This will enable learning to happen through computers and digital teachers will enable people to learn as per the comfort of their own schedules and at their own pace. Fast Forward to 2020: this has become a reality with online education, keeping the educational sector afloat amidst the COVID-19 pandemic.

Douglas Adams also wrote about an iPad-like device in his 1979 publication *The Hitchhiker's Guide to the Galaxy*. Through this device one could access all possible information available in the world, much like our very own smartphones and internet.

In *1984*, a dystopian work of science fiction published in 1949, George Orwell talks about the repercussions of mass surveillance being carried out on citizens by omnipresent cameras in a totalitarian government setup modeled on the lines of Nazi Germany. Orwell mentions 'telescreens' that are not unlike sensors, CCTV cameras, and facial recognition technology we have today. This was a good twenty years before the digital camera was actually invented. In today's context, aren't we too being closely monitored, with our digital footprint being carefully recorded and studied, and our social media feed crafted on the basis of it? If we search for hotels in Switzerland on google, we start getting advertisements on European vacation packages on our Facebook feed. Our search engines track all our searches, the YouTube videos we watch, and the hours we spend watching them. Isn't this too a gross invasion of privacy, no matter how accustomed we've become to it?

Recently, the global Pegasus scandal revealed that governments around the world used an Israeli spy software, now popularly called 'spyware', to illegally target and monitor hundreds of phones of journalists, human rights activists, and political opponents without the knowledge of the users. Isn't this also mass surveillance, and a version of Orwell's society?

HG Wells is often called 'The man who invented tomorrow'. He predicted the atomic bomb in 1915 in *The World Set Free*, with an astonishing degree of detail. In *Shape of Things to come,* he predicted airborne warfare. In his novel *The Invisible Man,* he used light refracting metamaterials, and this was almost a *century* before we actually learned about electromagnetic metamaterials! He even envisioned and introduced the world to genetic manipulation in *The Island of Dr. Moreau* in 1896.

Aldous Huxley's *Brave New World*, published in 1932, is a warning about the kind of world we are heading towards—too high on happiness to comprehend the depths of its depravity. Just that Huxley's brain conditioning chemicals have been replaced by the dopamine of smartphone and social media (this has been further discussed in the second chapter of this book).

Asimov's *Foundation Series* published from 1942-53 is based on a fictional area of study called 'Psychohistory', which studies the reactions of humans based on standard economic and social stimulus, effectively designed as a mathematical sociology to predict the behavior of thousands of people. Isn't this exactly what Big data and Artificial Intelligence do today? Predictive Mathematics has evolved by leaps and bounds, making us question if human beings are indeed that easily predictable.

It is evident that many of these authors got a lot right about the future. It can't all possibly be coincidence, or a stroke of luck? Perhaps the answer lies in the fact that most good science fiction has had its foundation in *good science* and a lot of successful sci-fi writers had a strong scientific background.

Arthur C. Clarke had degrees in math and physics. Before becoming a writer, he worked for the Royal Airforce. In his article *Extra-terrestrial relays*, published in 1945 in a British science magazine called *Wireless World*, Clarke described a plan for our modern system of geostationary communication satellites. It was fairly accurate, and as a tribute to him the altitude where such satellites lie is called "Clarke Orbit". Isaac Asimov had a Ph.D. in biochemistry, and HG Wells had a degree in biology.

Science fiction provides an overview of a plausible, futuristic society, looking at the big picture rather than the individual impacts of new technologies. While exercising their creative license to spin stories in make-believe worlds, authors must think through the multiple components of this make-believe world, including its logical fallacies and societal impact. This is in contrast to scientists, engineers, and technologists who usually solve one problem before moving on to the other, taking over (and sometimes destroying) the world one invention at a time, one innovation at a time. Science fiction gives us a synthesis of these innovations, in addition to streamlining what future innovations may look like. It provides a common understanding of the future in popular culture and imagination.

1.5 What about the Robots?

Two of the chapters in this book have been dedicated to Robots: chapter three and chapter five

Science fiction authors have always been fascinated with robots. The word 'Robot' was first used by Czech playwright Karl Capek in 1921 in a play called *Rossum's Universal Robots,* featuring mechanical machine-like men that worked in a factory in lieu of people. Eventually, these robots rebel against their masters. The origins of the word come from the Czech word 'robota', meaning forced labor.

By the 70s and 80s, robots were everywhere in popular culture—in kids' toys, various TV series, movies, comics, and music.

It was Russian Born American writer Isaac Asimov who gave the world a deeper peek into the concept of and the world of robots—more importantly, the role of robots in our world. He wrote extensively about robots, with several short stories, and over 15 books: *I, Robot* (1952), *The Rest of the Robots* (1964), *Robot Dreams* (1986), *Robots to Empire* (1985), *Robot Visions* (1990) to name a few.

Asimov devised three fundamental laws for robots:

1. A robot may not injure a human being or, through inaction, allow a human being to come to harm

2. A robot must obey the orders given to it by human beings except where such orders would conflict with the First Law.

3. A robot must protect its own existence as long as such protection doesn't conflict with the First or the Second Law.

With these laws Asimov gave a set of ethics that all robots must abide by, and explored the nuances and feasibility of his laws throughout his work. Notably, Asimov's three laws of Robots are a part of opening talks in practically every Robotics course at the undergraduate level.

His robots were portrayed as infinitely more advanced than the ones we currently have, and were capable of thinking and making decisions with a moral code (his laws) pre-programmed into them. However, it wasn't the functionality and features of his robots, but the light he depicted them in that changed modern media's perception of them. His robots weren't malicious machines that kill/ destroy humanity, or supercomputers to be afraid of. They were submissive and largely portrayed as helpful to the humans, somewhat more akin to the origin of the word robot behaving like a 'slave'. In one of his stories, they do take over humanity but only to prevent people from killing each other—another interesting take on his third law.

But some other authors question this slave-like status of robots, such as Philip K. Dick in *Do Androids Dream of Electric Sheep?* (1968). Here, robots demand their freedom and equal rights to their human counterparts, they are aggressive and violent when required. In general, Dick's future world view and robots' place in it was more pessimistic. He raises a difficult question—if robots are far superior to humans, why should they be lower in status or commanded by humans, only because they lack 'empathy' —something which wasn't granted to them by their human creators to begin with? Author Arthur C. Clarke also raises similar themes with his Robot Hal 9000 in *2001: A Space Odyssey*. Hal is an intelligent, thinking, creative robot, who can implement voice-based commands. However, Hal has to be unplugged after going mad on learning the truth about the concepts of life and death.

These authors raised pertinent issues about the role and rights of inorganic intelligent life, more than five decades before Saudi Arabia gave citizenship to Sophia—an intelligent humanoid robot developed in 2016 by Hong Kong-based Hanson Robotics.

The coming generations of robots are going to be far superior to humans in both physical and cognitive abilities, and we will be confronted by a plethora of questions on the framework within which these two races will coexist, and our perception of both in relation to one another.

Physicist Stephen Hawking warned us of the impending disaster that uncontrolled Artificial Intelligence can cause. In his book *Brief Answers to Big Questions*, published posthumously in 2018, Hawking narrates a dialogue using the word computer interchangeably with robot:

"People asked a computer, 'Is there a God?' And the computer said, 'There is now,' and fused the plug."

As per Hawking, it isn't that AI is out to destroy humans. But in going about achieving its objectives in an efficient manner, its prioritization will deeply impact our lives. For instance, I wonder that if the decision on whether or not to bring back Mark Watney at the cost of millions of dollars was left to a robot, in Andy Weir's 2011 novel *The Martian*, what would have been the outcome? Would a robot have thought that it was worth all the resources to save one life growing potatoes on Mars? (refer to chapter nine for more on Mars and space missions)

COVID-19 has increased our dependence on technology, and accelerated our move into the future. We are at the cusp of transitioning into the sort of future described in today's science fiction. Fiction forces us to stop and ask important questions with respect to AI and robots. Asimov, Clarke, and Dick are a good starting point to this discussion.

1.6 Sci-Fi, Policy & Politics

By articulating some threats that science poses or may pose, science fiction introduces scientific and technological innovations to public consciousness, making them an agenda in the political sphere. For example, over the past few decades sci-fi has highlighted the dangers of genetic engineering and AI, consequently sparking a public debate around them. Policymakers are therefore conscious of society's reaction to developments within these fields and work under that scrutiny. From *Brave New World* to *Jurassic Park*, science fiction has been cautioning us to be careful of what science can lead to. This increased public awareness and watchfulness pushes scientists to adopt a more humanistic perspective on technology.

Additionally, it is becoming critical for both people and politicians to get more interested in and examine the broader impact of new technology beyond the immediate gratification it offers. Unlike early inventions and innovations in aviation, automobiles, electricity, medicines, et cetera, many new technologies today are not solving critical problems but tinkering at the edges of the processes. Today's endeavors are often about improved efficiency and effectiveness using technology and process changes i.e., faster loan processing, driverless cars, better search and display algorithms, more user-generated content. Many of these help some consumer segments and entrepreneurs. But how much more productivity and how much more consumption do we really need? What will be their effect on society and employment? What should be the pace and path for these changes? Science fiction helps us be aware of this impact and showcases how solutions for non-existent problems often give rise to other unanticipated challenges.

My mother, an architect, says that it can take several years to get necessary approvals for constructing a building in a big city like London. This is because there are regulators analyzing the short and long-term impact of the building on the environment, heritage, and society. However, apps accessible to millions of unsupervised children around the world can be launched in a month. Can and should we extend this courtesy of in-depth analysis to virtual structures created by science and technology? Or is it sacrilege to interfere with the onward march of technology, free enterprise, and innovation?

We have no frameworks to assess the extent to which technology is impinging on privacy, human well-being, and market structures, nor do we have a framework regulating it. As Yuval Noah Harari points out in his book *Sapiens* (2014), human 'free-will' is a myth. It can be and is being manipulated by technology. Witness the

Cambridge Analytica scandal and how social media is now the most potent electioneering tool. And this is just the second decade of smartphones—imagine where we might be in 50 years! Add in the fact that technology is breeding natural global monopolies with capabilities and desire to destroy competition and subvert regulation.

Google's famous motto—'Don't be evil'—is an interesting reflection of the evolution of technology over recent decades. The motto was quietly buried in 2015 and replaced by a new one—'Do the right thing'—a belated recognition that Google perhaps had not, could not, or did not want to adhere to the earlier motto. A trillion-dollar company that touches the lives of billions quietly changing its motto from one which directed it to not do evil should be a matter of serious concern. 'Evil' is more objective than 'right'—do the right thing for whom, investors? Customers? Society? Quoting Arthur C. Clarke:

"I don't pretend we have all the answers. But the questions are certainly worth thinking about."

Today, we definitely have a problem—the deep disconnect between science and society, despite the inherent interlinkage between them—and something needs to be done about it. One source which does provide a framework to create scenarios of these impacts—science fiction—seem like a good place to begin.

This brings us to one of the most important discussions with respect to science, science fiction, and society—the relevance of ethical design.

1.7 Ethical Design, Science-Fiction, and Society

Science fiction reinforces the need for safeguards and socially acceptable guidelines that innovations in science and technology should follow and sparks public debate and opinions on what our future should look like. This sensitization is pivotal to arriving at a consensus, and a universally agreed set of norms on what ethical tech and ethical design within tech should entail.

Ethics necessitate never taking the technologies which are being imposed on society for granted, ethics require us to constantly question and investigate them. More importantly, they require us to alter and curtail them for the safety and security of human societies.

But why and how is ethical design relevant to the progress of science and technology?

Traditionally, science and technology set out with definite targets, and were, for the most part, successful in achieving them. These primary targets include examples such as creating an airplane, telecommunication, better transportation, medicines/vaccines for common diseases, et cetera. They had a common characteristic—they were easy to envision, were intuitive to human needs and fundamentally improved the quality of our lives. However, today, we no longer have such urgent, critical needs. The last two decades of advances in the internet and software have led to tremendous conveniences that have allowed us to work, communicate and access information better, faster and cheaper.

But we did not know that we wanted smart assistants such as Siri, Alexa, or Cortana till we got them. We did not know that we wanted to spend 30 minutes a day watching fifteen second videos. We do not know what we want till something interesting emerges from technologists 'messing around'. While we are more than happy to keep absorbing the gadgets thrust on us, *do we really need them?* And did we really demand them? We have entered a transitory phase where we are used to eagerly swallowing innovation after innovation as new technologies crop up, but have no concrete vision on what we want this innovation to look like in the long run.

Science fiction can help raise the right questions, help answer 'what do we want?', and inspire technologies (coming back to the smart voice assistants—Siri was broadly inspired by Gene Roddenberry's show *Star Trek* and Stanley Kubrick's *2001: A Space Odyssey*), but more importantly, it showcases the unexpected (and sometimes disastrous) outcomes that come of playing with technology whose potency we cannot grasp. It showcases the consequences of this lack of foresight, and the consequences of allowing unregulated agents in tech to operate independently and introduce technologies without considering their social implications, without being subject to any safeguards by society.

Take Facebook's 'like' button, for instance. 'Likes' run social media. They give validation, develop insecurities, augment social-media addictions, impact teenagers' perception of their self-worth, and social standing. There is dopamine to be found in a quick like, but this dopamine is dangerously addictive. Humans were never meant to deal with societal validation or rejection on the massive scale we are subject to today and base their behavior around chasing the former. We were never meant to account for so many opinions. The 'like' button and 'likes' have, arguably, done more harm than good. Most addictive substances that provide a dopamine rush are regulated around the world (think tobacco, alcohol, drugs) because we have had time to evaluate and think through the impact.

How does Big Tech justify altering the human psyche like this; what gives it the power to justify its decisions?

It can't. Nothing does. But Facebook never set out to create this toxic environment. They only wanted people to be able to appreciate others and feel appreciated. Individuals are not to be blamed for unprecedented societal outcomes of seemingly innocuous developments and features within science, but that does not absolve them of accountability.

Thus, there is a need not only for ethical design in tech, but for an emphasis on ethical design in tech.

Science-fiction bridges this gap, and this need for ethical design is a recurring theme found in most science fiction from the time of *Frankenstein*. Moreover, the very presence of books and stories that examine the impact that technology may have on society in the future makes us question what ethics we fundamentally agree on, and how to best go about implementing them. It also reinforces the *need* for us as a society to look for and debate those ethics. Science fiction and analyzing science fiction can thus play an important role in designing our future.

In March 2021, The US National Security Commission on AI said in its final report:

"AI systems will be used in pursuit of power. We fear AI tools will be the weapons of first resort in future conflicts. AI will not stay in the domain of superpowers or the realm of science fiction."

1.8 Conclusion: Science Fiction & the Future

While science fiction overestimated scientific advancement in certain areas, for example, intergalactic civilizations, it also underestimated these advancements in many cases. For instance, *Brave New World* is set in 2540, while at our current pace of innovation and development, we may achieve many of the technologies envisioned in it within the next century. As per Amara's law, which has been often cited in different forms, humans tend to overestimate the effect of a technology in the short run and underestimate its effect in the long term. While science has led to tremendous improvements in our health, comfort, and connectivity, the pace of change is now getting too fast for us to fathom, negotiate, and regulate,

As you will read in chapters eight and ten, *Snow Crash* in 1991 and *Ready Player One* in 2011 highlight the increasing power of tech entrepreneurs, the decline of nation-states, poor quality of physical lives and digital personas becoming more important than the real ones.

This is no longer fiction, recently Facebook has committed billions of dollars to its efforts to power a 'Metaverse', the term being taken directly from *Snow Crash*, a concept that will blur the difference between physical and augmented reality.

But we have not heeded the warnings or anticipated the scale and speed of impact. From 2008 to 2020, as human dependence on social media has increased, we have become 'products' at an unimaginable scale. Our psychology has become a product, sold by tech tycoons to advertisers looking to identify their target audience.

As mentioned earlier, climate change is another example of a large existential challenge which politicians and society have been unable to respond to. The timescale of the problem, the impact the potential solutions have on 'business as usual' and our 'normal'lifestyles, and the potential political backlash have made it incredibly hard to make any real progress on this front in spite of decades of warning. Many politicians called it fiction till 2018. Even today, we seem to be quite unprepared for what could come.

COVID-19 has been a painful reminder of how a lot can suddenly go wrong. It has also demonstrated that humans are capable of taking unprecedented decisions when the threat to life is imminent. But record forest fires globally, Canadian heat waves with temperatures soaring to 45 degrees Celsius and rainfall on Greenland's summit for first time in recorded history have not generated the same urgency of action around climate change. This demonstrates our inactivity and unconcern when the threat is less personal and a few years down the future. It is incredibly hard to shake the world out of its slumber, out of its inertia.

Brave New World and *1984* are revered today because they shook us to the core. They showed us disturbing societies which no one would want to live in, but perfectly logical societies—the kind which look out for collective good, guarantee happiness, the kind which everyone should want to live in. They showed us the importance of not always adhering to 'logic' and taking decisions that lead to a subjectively 'better' society and taught us to accept the inherent objectivity of 'good', to acknowledge our collective internal moral compass, and reminded us of an unspoken universal agreement on what is 'okay', what is 'good', and what isn't.

We need these reminders on what we want our future to look like, and what we do not want the future to look like. We need to have an approach to be able to take holistic view on role of science and not analyze each development in isolation. And that is what makes science fiction so worthy of our attention.

As we explore new, bold frontiers of science & technology, we need to be cautious that we don't end up with a frayed society.

As you read this book, I hope you gather valuable insights into how we should or should not structure our future, and what we need to be aware and alert of. At the very least, there is food for much needed thought.

Happy reading!

CHAPTER 2

BRAVE NEW WORLD

Aldous Huxley, 1932

Aldous Huxley's *Brave New World* has been one of the greatest dystopian books of the early 20[th]century. A savage satire that provided a fresh perspective on how science can potentially shape humanity, Huxley conjures a future World State set in the year 632 A.F (A.F here referring to 'After Ford'). Ford, the US auto magnate who conceived the assembly line for production, has replaced God as the reverential icon of civilization, and his methods of ensuring standardization and efficiency have been used to create a stable World State ruled by 10 'World Controllers.' Advances in reproductive science, sleep-learning, social conditioning, and recreational drugs are used to ensure all citizens are produced and brought up to strict specifications to live as happy consumers. Humans are bred and go through life in a standardized manner leaving no scope for individuality or extremes. They are conditioned to have a set of predefined choices (which are all met quite easily) leaving no room for unfulfilled desires, passion, or sadness.

The juxtaposition between a society worshipping a capitalist icon, yet foregoing individualism for an estranged notion of 'equality' and living under the control of a central authority is unique to Huxley. He's giving us everything the two extremes want—let there be a State looking out for the welfare of the people to provide standardized collective good, but let this good be happiness. Let this State ensure its citizens are perpetually happy, and get everything they want, with everything being based on logical principles. It seems like the perfect compromise, a utopian society.

The first part of the book is brilliantly conceived. The reader is introduced to the department of hatcheries and conditioning, where babies are created in labs with set intellectual/physical levels to be fit for predestined tasks and social roles or 'caste' (disturbingly similar to the caste system in India, where a child had a destiny laid out by society before it was born). The forecast manpower requirements for different jobs are provided to the hatcheries department who accordingly ensure the right amount of human supply for different professions. This perception of humans written in 1932 isn't very different from how humans are viewed today— as economic assets—which makes it all the more unnerving.

After careful fertilization, a foetus in a test tube is passed through an assembly line over nine months; carefully controlled nourishment and environmental inputs are provided to make it suitable for predefined tasks. For instance, mine workers in hot climates to helicopter pilots dealing with turbulence or assembly line workers with low intellectual aspirations. Once the babies are born, they are conditioned to think and behave in a certain manner through hypnopaedia (sleep learning) and other stimuli. This shapes their instincts and reinforces their belief in the defined social order appropriate for their caste and role in society. More brainy Alphas believe they have a great life, while the lowly Epsilons, menial workers, are also happy with their situation.

This prediction on lab-grown humans with specific traits is on the cusp of coming true, as the ethicality of genetic modification and human trials is under debate. But as society pushes towards 'better humans', and the technology to make them is available, why wouldn't anyone want their children to be healthier, stronger, and more attractive? The implications of normalizing the production of 'customized' children are what Huxley offers us, showing a scenario where human happiness and productivity are store-bought, but humans consequently become expendable; with individuality and idiosyncrasies becoming non-existent.

All adult citizens work efficiently during the day and play infantile games during their off time. Since women no longer give birth—the idea of childbirth is portrayed as a scandalous practice of the past—sex is purely recreational. Non-monogamous sex is expected and there are no repressed urges. There is no concept of family, love, marriage, or monogamy, *"everyone belongs to everyone"* is systematically drilled into citizens since childhood. It's a communist extreme, but with the hedonism that capitalism pursues. *Brave New World* rationalizes its society, as the 'Controller' explains in the book about the pre-modern world:

"The word was full of fathers - was therefore full of misery, full of mothers – therefore of every kind of perversion from sadisms to chastity; full of brothers, sisters, uncles, aunts – full of madness and suicide."

Long-term relationships, independent thought, loneliness, passion, or any behavior which goes against norms established for stable social order is frowned upon. The government provides a daily dose of Soma, a drug with no side effects which keeps citizens high and happy, to prevent people from tiring of their trivialized existence and searching for greater meaning. A society so high on happiness, can't comprehend the depth of its depravity.

In this 'perfect world' enters John Savage, a young man from an Indian reservation where tribes still live and reproduce as nature intended. Bred on a diet of Shakespeare and culture he is initially excited to experience this 'brave new world,' but finds the shallow and purposeless existence unbearable. The climax of the book is a dialogue between Savage and Mustapha Mond, the World Controller. Mond makes compelling arguments for his civilization, as it eliminated poverty, inequalities, conflicts, pain, and unhappiness. Savage, however, decides he prefers his old world back with its dirt, disease, poetry, danger, goodness, and sin. In short, free will. Quoting him:

"But I don't want comfort.
I am claiming the right to be unhappy."

While the moral higher ground the Savage takes is admirable, it is still subjective. What's scary is that Huxley made startlingly accurate predictions about the sort of choices that might develop with advancements in science, and the avenues it may open, and in a few decades, we may need to choose whether or not to head down those paths. As today's science is catching up to the 'fiction' many of these choices will emerge much sooner than 632 A.F. Creating our own 'brave new world' is theoretically possible, and in a few more decades will be as probable for society as not. We are at the precipice of choosing what weightage we give to ethics, and what ethics we, as a society, fundamentally agree on. Morals of a society cannot be detached from its political systems, and Huxley gives a beautiful demonstration of the same, while simultaneously warning us that we must eventually choose whether we identify with Savage or with Mond. Mond is the antagonist, but not necessarily the 'bad guy', and, to reiterate, does make many compelling arguments defending his society. Why wouldn't you want to be high and happy all the time?

On a digressing note, the historical context behind *Brave New World* is interesting. Written in 1931, three aspects seem to have been pivotal to shaping the book: the emerging liberal movement in England, rapid advancements in technology, and increasingly American influence in the world.

Firstly, liberal thought was gaining ground in England post first world war, exemplified by The Bloomsbury group that consisted of intellectuals like J.M. Keynes and

Virginia Woolfe. They advocated liberal, humanist values, one component of which was free love and sex. Their philosophy was at odds with the Victorian values and morality of the times. Huxley uses and extends these liberal thoughts to create a society in the *Brave New World* by deliberately taking on and decimating some of the icons of Victorian values. God is replaced by Ford, 'mother' is considered a shocking, scandalous word, indiscriminate sex is made into a virtue, and thrift is replaced with consumerism. As the audio conditioning tapes repeat thousands of times to sleeping children in the book. *"Ending is better than mending. The more stitches, the less riches"*. This contributed to the considerable shock value the book generated when it was published, to the extent of being banned in Ireland and Australia who somehow missed/ignored the inherent satire in the book. The savage reservations and civilization are also contrasted through the natural and the artificial: the reservation is made of naturally occurring products and colors, while civilization is manmade with apartments, bright lights, piped music, and perfume flowing in taps. Even colors in the book are mostly neon and pink with clothes made of viscose and acetate. Another instance where Huxley predicted what the future would and does look like.

Secondly, in the early 20th century, technology was transforming humanity and dramatically impacting day-to-day existence. From airplanes to radio, and television to medicine, there was a prevailing notion of technology solving all of man's problems. This belief is reflected in multiple utopian books, such as H.G. Wells' *Men like Gods* (1923), where men live in harmony within successful anarchy with no conflict. Huxley challenges this Wellsian utopian ideal by parodying it and taking scientific advances to their extremes to create a 'stable' society.

Thirdly, Huxley was also concerned that a consequence of the First World War will be the rise of America. He visited the US in 1926 and found its culture and mass consumerism as vulgar as he expected. Simultaneously, he expected that the 'future of America is the future of the world'. The World State in many ways is Huxley's projection of an American future, envisioning most buildings like skyscrapers, transportation in helicopter taxis, obsession with youth and consumption, 'feelies' that take Hollywood movies to the next level, and dresses with zippers. The book has a comical obsession with zippers—still a relatively new thing—almost as though it is an American invention meant to upset Victorian morality by making it so easy to get out of clothes. Lenina, the pneumatic female lead sleeps in zippyjamas. John ponders that he just has to *take hold of the zipper at her neck and give one strong pull'*. The female undergarment is called *'zippicamiknicks'*. *'Zip, zip she gets out of her bell-bottomed trousers'*. It goes on...

Ninety years from when it was written, it's fascinating to explore the book's relevance today and how its blueprint has unfolded. Liberal values have continued to grow while monogamy has declined. Social apps and smartphones keep people continually stimulated, and warp their mindsets. Huxley relied on test-tube babies, chemicals, and sound and light effects to condition the brain. He couldn't have thought of the technological capabilities of smartphones, social networks, and AI to influence the human brain and redirect our free will. As Yuval Noah Hariri's *Sapiens* suggests, we may not need 600 years to learn to master and condition the human brain. Nazi Germany and the USSR tried to create their own models of Utopias, although both ultimately failed. American influence in the world did indeed grow considerably from 1930 till very recently, now finding somewhat of a counter in Asian culture and ideals.

Coming back to the core focus of the book, Huxley does not totally denigrate this new society. There is an undercurrent that shallow happiness and fulfilment without free will and emotions is an ignoble way to live, and readers do get the satire. But the author's ambivalence about the morality of such a society is evident through Mustapha Mond in his conversations with Savage and Helmholtz. In fact, Huxley in a foreword written in 1946 says that he should have given a 3rd choice between 'an insane life in Utopia' or 'the life of a primitive in an Indian village'.

And while I personally still see it as a vacuous society, with banal existence, high on pursuit of inconsequential happiness, I don't doubt that many will disagree. Huxley poses an alluring, timeless question:

Looking at the suffering, poverty, and depression in our society, will it really be such a bad solution to give up free will for a secure, happy, and disease-free existence?

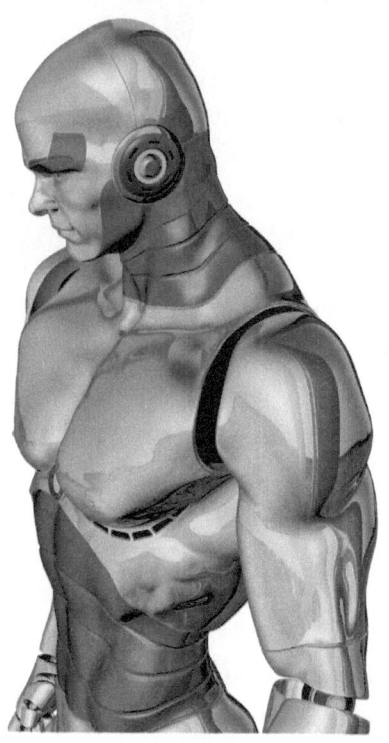

CHAPTER 3

I, ROBOT

Isaac Asimov, 1950

Isaac Asimov's *I, Robot* is the first book of his Robot series, and is easily one of science fiction's most influential works, establishing the often-cited Three Laws of Robotics.

It is a collection of nine short stories that were written between 1940 and 1950 with the plots set in time frames between 1996 to 2052, when rapid advancements are projected to take place in robotics and artificial intelligence.

Asimov envisions a future where robots are commonplace and explores the ethicality—both of having robots, and the ethical code they might follow—through his laws. He doesn't present his laws as infallible, but rather makes a genuine effort to analyze the loopholes within them, and paradoxes that might present themselves.

This evaluation of highly intelligent and capable non-sentient beings through a moral and psychological perspective is particularly interesting today, as we've only recently begun debating ethicality in and of AI (Artificial Intelligence), and dismantling and understanding its biases. While Asimov doesn't directly tackle biases, he highlights the importance of safeguards against man-made higher intelligence and the need for clearly established and universally agreed ground rules pertaining to them. *I, Robot* strings together multiple anecdotes involving robots of various degrees of intelligence, tracing their interactions with engineers Gregory Powell and Mike Donovan, and robopsychologist (as the term suggests, a psychologist whose expertise lies in the robot psyche) Dr. Susan Calvin.

The basic plot of the book that links the nine stories revolves around a reporter's interview with a seventy-five year old Dr. Calvin as she recollects the dysfunctional robots, complications of their 'positronic' brains, and the inherent problems of human-robot interactions. The stories center on the challenges that arise in the realm of ethical technology design and programming.

The three laws of robotics, proposed by Asimov are as follows:

1. A robot may not injure a human being, or, through inaction, allow a human being to come to harm.
2. A robot must obey the orders given to it by human beings except where such orders would conflict with the First Law.
3. A robot must protect its own existence as long as such protection does not conflict with the First or Second Law.

Thus, we have a framework that seems perfect on the surface, perfect to control and guide robot behavior, and is meticulously examined and dissected through a five-part series. Furthermore, Asimov doesn't present a society that wholeheartedly accepts the machines and supports their seamless integration into human lifestyles. On the contrary, he gives us humans as skeptical and wary of robots as we are today, with some people viewing them as abominations to be avoided as far as possible. There's plenty of conflict between US Robots and Mechanical Men Inc. (the quintessential innovative capitalists), the government, and the people. Further, there's conflict within the robot manufacturers, as the organization heads and faculty have different perspectives on how they view the robots. Calvin views them as a cleaner, better breed than humans, making more rational and ethical decisions due to their moral code of conduct. She's characterized by an icy demeanor and a 'mechanical' heart. Powell and Donovan are slightly more comical, the good-natured engineers who get their hands dirty and save the day. Overshadowing them are the directors and officers who view the machines as just that—machines. The delicate interplay of these conflicting viewpoints has a startlingly accurate depiction of modern society's perspective on artificial intelligence—conflicting. We still have capitalists concerned purely with monetizing off complex algorithms (present-day US Robots and Mechanical Men Inc.), scientists fascinated by the ethicality and technicalities of artificial intelligence and the robots' 'positronic' brains (a shade closer to

Dr. Calvin), those who scrunch their brows at terms as convoluted and intimidating as 'artificial intelligence ' (similar to the reaction common people have in Asimov's world), and starry-eyed engineers and 'techies' who simply want to be there (the likes of Powell and Donovan). For a book published in 1950, Asimov's novel take on robots gets a lot right. It was prescient about identifying issues that we have started grappling with only in this century.

An interesting avenue *I, Robot* went down was Calvin's perspective on robots as the morally superior species. One of the stories revolves around an electoral election, with two politicians vying for the mayor's post, Francis Quinn and Stephen Byerley. Quinn suspects that Byerley isn't fully human, and goes to Dr. Calvin for help, claiming that Byerley has never been witnessed eating or sleeping—two basic human needs. Calvin provides a counter to Quinn's suspicions, asserting that Byerley may as well be a robot if he adheres to the rigid laws of robotics, or he "may simply be a very good man" since the universal regulations framed for the robots happen to be, quoting Calvin, "the essential guiding principles of a good many of the world's ethical systems". It is the inhumanity of the robots (paradoxically created by humans themselves), that renders them incapable of disobeying the laws, and they have no selfish or self-serving motives. After all, there is no universal moral code of conduct that humans have to abide by. Quinn's suspicions are testament to our perception of politicians—selfish and corrupt; it was, after all, Byerley's perfection as a politician that attracted Quinn's attention. So wouldn't robots, or any intelligent beings who have a variation of Asimov's laws pre-programmed into them, make better politicians, better law-makers than humans? They'd be unbiased, selfless, and would always adhere to their pre-programmed ethical ground rules. It's a bitter truth, but one we must eventually confront. Or rather, a question that we may one day need to address and come to terms with.

Asimov's laws are ambiguous, his verdict on society's treatment of robots uncertain, but he unapologetically showcases, and in most cases resolves, this ambiguity. The evolution of both robots and their functions is chronological. The first story in I, Robot is of Robbie, a non-vocal nursemaid (robot) made in 1996, and explores human attachment to robots. He babysits Gloria, a human girl, who's devastated when her parents take Robbie away from her as they fear that she's too attached to him. Gloria and Robbie eventually reunite, and we're left with a 'happy ending'. The fifth story is about Herbie, a mind-reading robot made in 2021 who faces an ethical dilemma while interacting with humans due to the first law and lies to them in order to not hurt their feelings. Herbie eventually has a breakdown, when in the presence of people with conflicting interests, as a statement that'll please the first will hurt the other and vice-versa.

The last story, *The Evitable Conflict*, set in 1952, projects a world dominated by supercomputers. The Machines strive for control, only to save humanity, for the betterment of the human community. The robots have an organic goal to achieve (the protection of the human race), but their complete sovereignty over the humans also implies that at the end of the day, they are far superior in intellect than their creators. Dr. Calvin, previously shown as someone inhumanly fond of machines, can also be witnessed claiming "Mankind has lost its own say in its future" towards the end of the novel, but reiterates her previous statements in the beginning of the novel, by asserting that it is better this way since robots happen to be the only things which "stand between mankind and destruction". To err is to be human. With Machines in control, large-scale disruptions (wars, economic turmoil, etc.) will be avoidable–evitable.

"Only the Machines, from now on, are inevitable!"

The stories were written at a time when World War II was ravaging countries and continents, and the atomic bomb had been unleashed. Asimov's implicit emphasis upon the possibility of the creation of a superior being, by ascribing idealized human standards to them, is also perhaps an expression of hope for global peace.

Thus, Asimov touched upon topics with differing degrees of complexity, crafting a thought-provoking structured narrative, and greatly altered society's perception of robots and artificial intelligence. He skillfully tackles the 'Frankenstein complex'—the inherent fear humans have that their creations will turn against them—which pop culture perpetuates when talking of automatons and androids (think *Transformers*) by providing a framework that regulates robot behavior, and elaborating on its nuances. While achieving the level of self-awareness and intelligence Asimov's robots had is still far off, the Robot series firmly reinforces the need for and importance of ethics in AI. 'Ethics' being a subjective, morally grey area is, of course, another debate.

CHAPTER 4

THE MOON IS A
HARSH MISTRESS

Robert A. Heinlein, 1966

Robert Anson Heinlein was a contemporary of Isaac
Asimov and Arthur C. Clarke, the trio termed the "Big
Three" of English science fiction authors. Through his
narratives, Heinlein captures and explores social messages
such as the significance of individual liberty, the impact
of organized religion on society, and the societal urge
to repress rebellious and revolutionary ideas and actions.
Heinlein provides an overview of current socio-political
frameworks and speculates on how they may change in
the future, in response to developments in the fields of
science and technology.

The Moon is a Harsh Mistress was published in 1966
and is considered one of Heinlein's finest works, winning
the Hugo Award and being nominated for the Nebula
Award. Set in the years 2075-76, it showcases a future
where a Lunar community is organizing a rebellion
against the absentee rule from Earth. Heinlein creates
a futuristic society and demonstrates how science and

technology may shape the future, how some socio-cultural practices can be organized differently, and what a new libertarian society could look like, particularly as humans make a fresh start at colonies in space.

The plot is as follows:

The Moon has been inhabited, starting off as a penal colony comprising criminals and other political exiles, and growing organically to house over 3 million people across various nationalities, cultures, and languages. From Earth's perspective, the primary function of the Moon is to provide wheat for its burgeoning population. Due to the moon's harsh environment, people live and conduct all their activities, including agriculture, underground. Important lunar systems and infrastructure are controlled by a powerful central supercomputer, Mike.

Mannie, our narrator, is a computer technician who discovers that Mike is developing self-awareness as well as a sense of humor, and they gradually become friends. Other key characters in the book include Professor Bernardo de La Paz, a self-proclaimed 'Rational Anarchist' with strong libertarian beliefs, and Wyoming ("Wyoh") Knott, a political agitator from the colony of Hong Kong Luna.

The novel is divided into three parts—Book 1: That Dinkum Thinkum, Book 2: A Rabble in Arms, and Book 3: TANSTAAFL! respectively.

Book 1 describes the Lunar society, one where citizens are under the control of 'The Authority' of Earth, run by a Warden. The key objective of the Authority is to ensure timely wheat shipments to Earth. Beyond that, it is a fairly anarchist and self-regulated society of 'Loonies'. The dystopia emerging out of the novel is of an economic sort – the citizens of Luna are coerced to engage in business with the Authority. Professor de la Paz summarizes this system of exploitation:

"It strikes at the most basic human right, the right to bargain in a free marketplace."

The novel utilizes its lunar setting to explore the nuances of an inter-planetary revolt and weaves several genres of dystopian fiction, philosophy, and science fiction. De la Paz convinces other residents that the Moon needs to stop exporting wheat to earth, as it is depleting its precious and limited ice-based water resources. Together, De la Paz, Mannie, Wyoh and Mike start a revolution and seize power from the Lunar Authority. Mike plays a central role by controlling all communications within the Moon, and to Earth. This underlying theme of self-aware, sentient artificial intelligence which goes as far as to rebel against the purpose it was built for is disconcerting, and while improbable (for now, thankfully), does warn us of the risks of not imposing rigid safeguards against computers and artificial intelligence in general. Quoting Stan Lee's Peter Parker, "with great power comes great responsibility". We, as humans, have a responsibility towards both society and ourselves to ensure that we don't arbitrarily allow man-made creations or artificial intelligence of any sort to have full control over any human systems or structures unless we can guarantee that that power won't be misused. Heinlein's depiction of AI, with its astounding self-awareness and human characteristics, may still be far off, but the warning echoing through the novel is sound.

In Book 2, A Rabble of Arms, Luna declares its independence on July 4, 2076, the 300th anniversary of the US Declaration of Independence, and crafts a declaration borrowed from the US declaration. Whilst Luna as a penal colony reminded me of how Australia was initially populated, the lunar society also borrows in many ways from the US—a new country that is a melting pot of cultures and has relatively higher norms for individual liberty.

Mannie and Professor then visit Earth to negotiate independence for the Moon. This trip is interesting as it outlines the political climate of the Earth in 2075. The world is governed by a Federation and the major nation-states under it are Europe, North America, China, and India. Interestingly, China and North America are portrayed as the most powerful and influential of the lot with the former having annexed a huge chunk of East Asia along with Australia. India has been portrayed as an over-populated nation.

It is uncanny how Heinlein predicted China and India's role as the important future powers, even though they were both very poor and populous nations when the book was penned in 1966. Today we see China expanding its influence across Asia and challenging the American role as the 'Superpower', with India too starting to rise.

It is interesting that most science fiction books and movies that deal with space colonization tend to create a single governing federation on Earth. But is it really likely that in a scientifically advanced society, we will have a single body governing the Earth and all its diverse Nations.

However, the negotiations are unsuccessful and Mannie and Prof return to Moon. They form a new government on Moon, which they are elected to lead, with some help from Mike. Mike creates an imaginary character, Adam Selene, who appears to the masses only on a video screen and is positioned as the leader of the revolution.

Book 3, is titled 'TANSTAAFL!' after the political slogan, "There Ain't No Such Thing As a Free Lunch". This slogan gained so much popularity post the book that the Libertarian Party appropriated the same as their slogan a decade after the novel by Heinlein was published in the market.

The Federation of Earth attacks the moon but the attack is thwarted by Luna. Luna attacks the earth by firing large projectiles in uninhabited areas. They are pulled to earth due to gravity and have destructive power equivalent to an atomic bomb. Mike coordinates the firing of these projectiles by making precise calculations and eventually, Luna is granted its freedom.

The book presents a believable image of the future and lunar colonization, overall a story that feels futuristic but possible by 2075. The advanced technology presented in the novel seems scientifically precise and unambiguous. The process of hurling lunar rocks at Earth (intended to be destructive and calamitous), while somewhat primitive, seems to be scientifically feasible and impactful. At the same time, the book depicts a symbiotic and almost intimate bond that could exist in the future between humans and computers and the humanization of artificial intelligence.

Mannie and Mike share a special relationship and Mike's nickname for Mannie is "Man". Mannie has also been positioned as a 'part-machine' as he had lost his left arm in an unfortunate mining accident and was provided with mechanical robot-like prosthetics designed as artificial hands. Mannie confesses in a chapter that he comprehends machinery in a finer manner than he understands humanity.

Mike has been portrayed as a supercomputer who has the option to turn rogue (for instance, HAL-9000 computer in Arthur C. Clarke's *2001: A Space Odyssey*) but the novel churns out a rather happy ending, without the occurrence of any major apocalypse.

While set in 2075 against the backdrop of the moon, the book does not overtly focus on technological advancements. There are regular space shuttles between Earth and Moon and there is Mike, the supercomputer. But beyond that Heinlein keeps the focus on politics and society.

This book is an ode to Libertarian ideals as it shows how a society can self-govern and regulate itself with good outcomes. Professor Bernardo de La Paz, the 'Rational Anarchist', upholds the belief that the existence of the government is to deceive the citizens. However, he realizes that some form of government is necessary, regardless of its heinous flaws. When Professor de la Paz and his beliefs are disputed by Wyoh, the former asserts,

"In terms of morals there is no such thing as a 'state'. Just men. Individuals. Each responsible for his own acts. I am free, no matter what rules surround me. If I find them tolerable, I tolerate them; if I find them too obnoxious, I break them. I am free because I know that I alone am morally responsible for everything that I do."

The Professor claims that there is no theft or robbery which occurs on Luna and when he is engaging in a conversation with Stu, Mannie claims that "there is no rape on Luna", that "every man pays his dues" and "all acts of violence are justified". The fact that this colony of Loonies" started out as a penal colony, renders a further utopian touch to the setting. Owing to the presence of lower surface gravity than that of the Earth, inhabitants who stay back for a period of more than six months, go through "irreversible physiological changes" and can never live comfortably under 'normal gravity', thus making a return to Earth, an unreasonable attempt. This aids in providing a perfect setting for a scientifically advanced society unburdened and unchallenged by the rules of Earth.

Apart from the heavy Libertarian undercurrent, the book also seems to be influenced by Malthusian concepts as Earth is shown to be running out of space and food grains by 2075, with its ever-growing population. Thankfully, it is now apparent that this concern was misplaced, and Earth's population will stabilize by end of 21st century.

In terms of the Lunar Society, Heinlein presents a world with no racial, cultural, or religious biases. As the Loonies population grew on the moon, they developed their own social structures and customs. This may not seem very surprising now, but in 1966 biases around color and race were widespread. The language created to serve the purpose of being the 'lunar dialect' has its base in English with a coalescence of Russian, American, and Australian dialects. The language follows the American style of diction, but has Russian syntax —an evident instance being the second line from the novel, "I see also is to be mass meeting tonight."

The marriages which take place in the lunar society are mostly polyandrous. The book also introduces the idea of 'line-marriage', where additional spouses keep getting added so the marriage never ends. Several characters, including Mannie, indulge in the practice of the same.

However, the depiction of women in lunar society is a bit confusing. There are fewer women than men on Luna, so there is a lot of respect and empowerment for them. There is a clear acknowledgment that women's bodies are their own and their choices about sex, marriage, and children rest with them alone. As mentioned earlier, women often have multiple husbands and run a large line of married families. It is understandable that in 1966, even this would have been a significant change in the perception and depiction of women. At the same time women, including Wyoh, are not given any substantive roles in the revolution or the government and their power seems to be confined more to their families. The stated fact that the gender imbalance of 2:1 leads to respect for women, makes it seem more a matter of survival for Loonie society and dilutes the book's positioning of gender equality.

Heinlein curates a balanced futuristic narrative and a detailed and meticulous account of the society prevailing on the moon and the relationship between man and machine. The plot structure echoes political sophistication and embodies several unique stylistic elements to enhance the suspense and thrill. While set in a future with certain technological advancements, the book focuses primarily on society and politics, a significant departure from the typical science fiction of the time. Science and technology are not positioned as the cause of any major challenges to society but are primarily used to justify and enable a society on the moon. The core of the book is then free to focus on how such a society and culture can be different compared to the earth.

Leigh Kimmel appreciated Heinlein's use of

"colloquial language... an extrapolated lunar creole that has arisen from the forced intersection of multiple cultures and languages in the lunar penal colonies, the protagonist's disability, the frank treatment of alternative family structures, and the computer which suddenly wakes up to full artificial intelligence, but rather than becoming a Monster that threatens human society and must be destroyed as the primary quest of the story, instead befriends the protagonist and seeks to become ever more human, a sort of digital

Pinocchio."

Quoting Adam Roberts,

"It is really quite hard to respond to this masterful book, except by engaging with its political content; and yet we need to make the effort to see past the ideological to the formal and thematic if we are fully to appreciate the splendor of Heinlein's achievement here".

To conclude, it is a fun book to read with a great plotline and likable characters. Towards the end, we find ourselves rooting for the Loonies, and their newfound freedom and government.

Though serious colonization of space is still far-off, whenever it happens, it will create new questions and opportunities around the socio-political structures that will govern such societies and also their relationship with Earth. Another instance where we can't ignore the deep interlinkage between scientific innovation and society, or view them independently of each other.

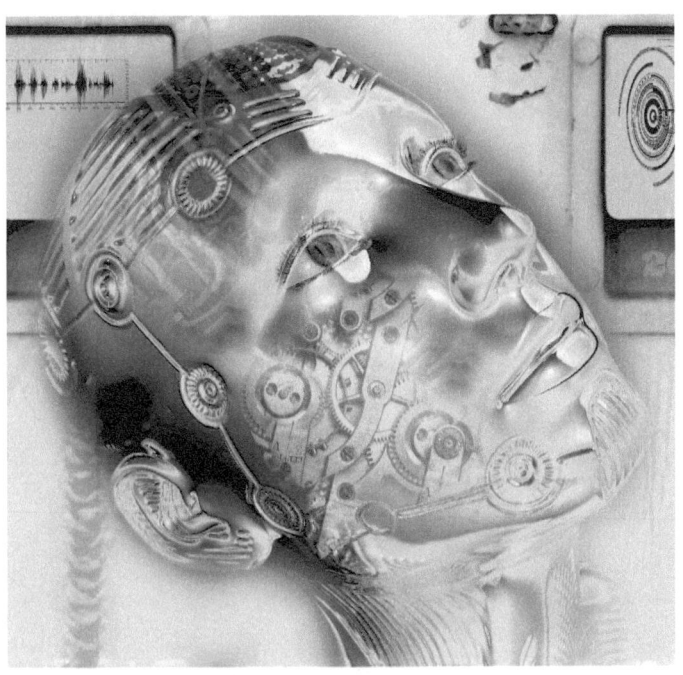

CHAPTER 5

Do Androids Dream of Electric Sheep?

Philip K. Dick, 1968

Philip K. Dick's novel *Do Androids Dream of Electric sheep?* was published in 1968, and read by me in 2019. It was made into the 1982 classic movie *Blade Runner* which catapulted the book and its author into fame. The 'androids' mentioned in the title are not exactly what we in the 21st century would assume. For us, the android points to the android operating system for smartphones. Dick refers to the older, original meaning of the term 'android', an electric entity or robot possessing human appearance and not having the capacity for human emotion, or experiencing only a very limited range of them.

In Dick's novel, this range excludes one of the most important, humane emotions which humans possess: empathy. This lack of empathy makes androids a danger to humanity, and there exist bounty hunters like Rick

Deckard, our protagonist, assigned to 'retire' these andys (androids). The title of the novel poses a philosophical question which muses on the extent of the androids' inhumanity: do androids have the ability to dream, and if yes then what do they dream of? Do they too wish to own an actual living animal, as Deckard does?

The novel, set in a dystopian 1992 (2021 in later editions), features a destroyed Earth that has been made unfit for living by the World War Terminus. World War Terminus left a destructive radioactive ash in its wake, damaging the homes, lives, and mental faculties of humans and other living organisms. Most humans have migrated to 'off-earth' colonies like on Mars, where they get androids to work as slaves.

Fortunately, we have entered the 21st century without serious threats of a nuclear war, or another world war. But climate change is still a very real threat, and artificial intelligence as well as space travel are slowly gaining ground. The dystopia Dick describes may very well be one that we're working towards, and within it Dick explores the nuances of the relationship between humans, and human-manufactured intelligence. Of course, the degree of intelligence Dick's androids possess is a long way from what our artificial intelligence looks like, but he draws light to how much we value human emotions, and empathy—something which no amount of intellectual prowess or the physical superiority of robots can replace.

Dick's androids look like humans and sometimes escape to the earth. There are teams of bounty hunters on earth to hunt down and destroy these androids. However, distinguishing androids from humans gets harder and harder as androids get smarter, and special tests are designed for the same.

The story traces a day in the life of the bounty hunter Rick Deckard and his mission of the day: to retire six Nexus-6 type androids. The mission is particularly difficult because Nexus-6 type androids, as told by his boss and the initial bounty hunter who pulls out of this mission because he's severely injured, are very intelligent creatures. The six Andys are sharp enough to impersonate humans from various walks of life and cheat to survive. In the end, when Dick manages to retire all six of these creatures, his basic morality and outlook of life has changed, and he no longer feels the same as before with respect to humans, androids, and their respective behaviour.

However, the book is not really about the battle between a human and androids. Dick's sci-fi also elaborates on various topics and themes, of a society where humans and technology are getting closely intertwined. As mentioned earlier, an important characteristic differentiating humans from androids is the capacity for empathy. Hence, there is a big cultural push amongst humans to cultivate greater empathy. Two approaches for humans to get more empathetic, as portrayed in the book, are to acquire animal pets and connect to empathy boxes.

Deckard owns an electric sheep. His dream is to own a real animal, in lieu of his current one (he did own a real sheep before it died, and was converted into the electric one). Animals are killed by the radioactive ashes, and those left are highly coveted rare entities carrying significant prestige. Big animals, like sheep or goats, are higher in the social hierarchy of perceptions of prestige than rabbits, squirrels, rats, or cats. Deckard too wants to buy an animal not for his love of animals, but to enhance his own self-worth and pride.

"Owning and maintaining a fraud had a way of gradually demoralizing one. And yet from a social standpoint, it had to be done, given the absence of the real article."

Not owning an actual animal, as Deckard ponders, has "sapped his morale" and he tries to overcome this by buying a goat, the ultimate luxury a human can have.

When he receives the bounty for killing the first three andys, he buys an actual goat with his entire bounty money. But it doesn't bring him the peace that he was looking for:

"The expense, the contractual indebtedness, appalled him; he found himself shaking. But I had to do it, he said to himself....I have to get my confidence, my faith in myself and my abilities, back. Or I won't keep my job".

This desire to have the 'next big thing', to go from an electric sheep to a real goat and still be dissatisfied and want something better, isn't unique to Deckard, it's something we all have. There is an inherent human desire to possess things, as is evident all around us—be it a newly launched iPhone, with three cameras instead of the older one with two, or a new video game with the promise of a better interface. We know that we don't necessarily need an update, but we want it nevertheless.

We are led to believe that what we seek is unique, and it is imperative that we possess it. Just like a real animal is in the post-World War Terminus of the novel's dystopia. And just like Deckard, we too, place our self-worth in these things and items. And often the joy of ownership is short-lived before we start craving the next big thing.

This principle extends further to developments in technology as a whole. We relentlessly pursue greater efficiency, nicer interfaces, exciting gadgets, without pausing to wonder why we are. Without picturing where this will take us, and do we really want to go there?

Rather than having a clear vision for the future and working in accordance with achieving it, we continuously alter our vision of the future depending on whatever new technologies spring up. While being open-minded and flexible is a good thing, how long can we keep up with this disconnect? A better approach would be aligning technological development with the needs of society, while maintaining enough flexibility to accommodate new developments in science and technology, and incorporate them into a larger, clearly defined framework.

But I digress.

Coming back to Dick, and Deckard, another interesting tool to cultivate increased empathy are the 'empathy boxes' present in human houses. This empathy box offers various moods, depression included, to dial and experience. It's interesting that a book published in 1968 was able to visualize that humans would need mechanical devices in order to feel connected to the world.

That humans would become so detached, they would need external support to control their emotions, and connect.

Another way of looking at it would be that humans wanted such complete control over themselves, in this quest for efficiency and productivity, they developed a means to do so.

Even today we're seeing a version of the same concept—of this overdependence on technology and quest to be 'better'—an ever-growing reliance on digital devices and a virtual world has left us socially paralyzed to the extent that we can barely function offline.

This has been augmented with the pandemic, and our transition into giving more importance to our virtual personas than our real ones has been accelerated. Deckard's wife using her 'mood organ' to share her joys with others, to seek pleasure from others, to connect, isn't very different from what we do today. We too share our lives and emotions with others through social media, and derive pleasure and societal validation through likes and comments.

Coming back to androids. They're designed to be human slaves on Mars, but hunted and retired if found on Earth. They are positioned as a necessary and dangerous offspring of technology invented by humans. The novel offers a bleak representation of the way androids function. They are soulless beings with an outer shell that represents a human. But as Deckard realizes eventually, it's always going to be humans versus androids. Androids lack the empathy to survive and care for each other the way humans do, or at least are expected to do. Despite highlighting this divide between humans and androids, Dick doesn't hesitate to show that humans too can be apathetic to fellow human beings. This is demonstrated through the character of Isidore, who is classified as a 'chickenhead'—a derogatory term used to refer to people whose mental faculties have been altered and damaged by the radioactive ashes on planet earth. These chicken heads are 'special' and are treated as second-class citizens, with a bare minimum in terms of living standards. Isidore, in many regards, is mistreated by humans, because he is different from them, and perceived as dumb. This lack of empathy shown by humans even among their own race makes them hypocrites when it comes to expounding the ideas of empathy to androids. Through the inner workings of Isidore's mind, Dick asks the question of the standards humanity can stoop to when not examined or challenged, where we target one another, creating a hostile environment for ourselves as well as other beings. Empathy, while highly rated and the key differentiator vis a vis the machines is not necessarily an arena human beings themselves flourish in.

This crippling effect of technology imagined in late 1960 through this book has been substantially amplified over time. The book puts forth the question of how much can a human claim to be human if they need an empty box or a mood organ to dial in their mood choice and to feel something by fiddling around a machine. Our mood organs are our phones, tablets and TVs, all of us seeking mechanical help in the end. Thus, the argument that humans are superior to specials like Isidore and androids because they possess emotions like empathy, is a flimsy one. After all, Isidore is shown to have more empathy for animals or living beings than both Deckard, a human, and the three androids Isidore is sheltering, who torture a living spider by cutting off its four limbs, mutilating it for their amusement. Isidore is in pain, mentally, when he sees this happen and releases the spider when he runs into Deckard, who asks him as to why he released the spider since he would have gotten a tidy sum of money for it. While Deckard sees animals and any other living organic entity as a means of social status or money, Isidore doesn't; Isidore is also different from androids because he doesn't envy the sacrosanct value placed on animals, as the androids do. Isidore is better at feeling emotions than his superior human counterpart and the empty and abstract androids.

However, the entirety of any human existence survives through shared empathy, along with the individualism of a person. This is shown through the way Deckard, after finishing off his mission of killing all the andys (a personal record for him), also comes to terms with his identity as a bounty hunter and animal owner. The goat that Deckard buys is killed off by his android lover, however, he eventually realizes that he is far better off with his original electric sheep.

Towards the end of the novel, Deckard as well as the readers realize that the androids and humans are not that different: both want to survive and have an inherent desire to let the life intended for them unravel, the life they think they deserve. The only difference is their lifespan (for androids it's 4 years) and the inherent empathy for the value of fellow human beings as well as the animals. The androids can easily kill one another, while humans like Deckard are a bit skeptical about it, if not downright unable to go through it.

The lessons that Dick provides us through Do Androids Dream of Electric Sheep? are many, one of them being the true nature of a living entity. As Deckard realizes

"The electric things have their lives, too. Paltry as those lives are."

Insects, humans, specials and androids, all have lives, electric and organic. They just don't fit into the other's definition of what it means to be truly alive. This is a very pertinent question as we think of forthcoming advances in biomedical electronics and genetics. The expectation of the rise of machines and superhumans along with the declining economic value of humans will lead us to some new set of existential questions. How will we define what is human, who can be retired, whose life will be at a premium, and why?

Through this book, Philip K. Dick leaves us with a profound question: what does it mean to live? What is a human life and how should it be lived to ensure fulfillment and who is the judge of whether life should survive or not?

We must ask ourselves, and answer these questions before experimenting with genetic engineering, or with higher degrees of artificial intelligence. We don't want to find ourselves in the positions Deckard does, to not have a clear idea on how we perceive future 'androids' (or variations of the same) while coexisting with them.

As far as the title of the book is concerned, no, the androids don't dream of electric sheep. They can't even understand why humans do. That's what makes them androids and maybe that's what makes Deckard and maybe us, humans.

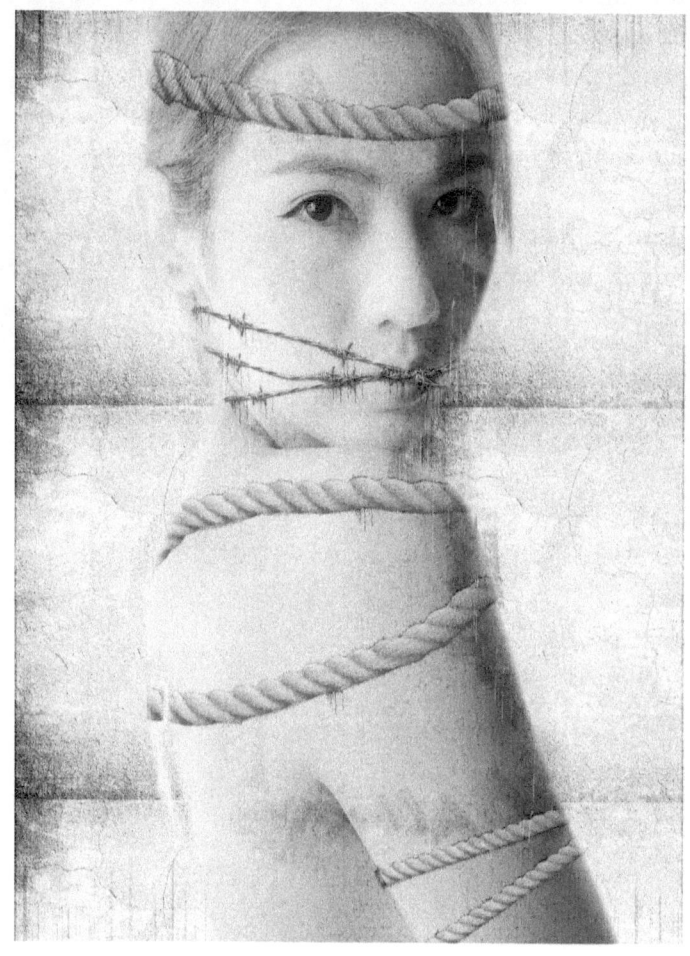

CHAPTER 6

THE HANDMAID'S TALE

Margaret Atwood, 1985

While in conversation with my aunt, our focus shifted towards books. She'd recommended *The Handmaid's Tale* to me, and after reading it I was eager to get her inputs on it (and bombard her with my opinions). Surprisingly, she showed no pleasure in indulging me despite my persistent pitiful attempts to steer talk towards a book that I hadn't been able to (and probably never will) get over.

"Oh, the writing was good, but I didn't like it very much", she sighed.

"Bears an uncanny resemblance to present day and how we as a society are, just removed all the sugarcoating. It felt too real. Another coffee, please"

Written in 1985 and read in 2020, Margaret Atwood's *The Handmaid's Tale* is a risqué venture into dystopia.

The setting – At a time when harmful effects of radiation and chemicals in the environment have severely reduced fertility and birth rates, The United States government is overthrown and replaced by a military dictatorship known as the Republic of Gilead. The new regime organizes the society into new classes as per their needs to increase the birth-rate. People are segregated into a hierarchical class system and dress codes are imposed to reinforce the 'place' of everyone in society. Women's rights are curtailed and they are not allowed to have a voice, read or work, they remain confined within the four walls of the house. A new class of fertile women has been created who are called 'Handmaids'. They are trained and coached to live a certain kind of life where their primary 'job' is to reproduce. They wear long red cloaks with white hats covering their face on the sides and big boots. Any proven fertile woman who has been a gender offender or married a divorced man is condemned to live the life of a handmaid. They are separated from their children who are adopted by high-class families with infertile women. Handmaids are then trained at designated centers by an overseeing 'aunt'. Senior class men with infertile wives are allowed to have Handmaids as a status symbol who will bear children for them. Once the Handmaid has a child, she nurses the newborn and after being weaned off, the Handmaid moves to the next 'job' to another upper-class household to continue producing children. She is treated as per the mistress of the household's wishes, and is considered "untouchable".

Atwood thus gives us a 'blast into the past', describing a futuristic society wherein women have virtually no rights. Our protagonist is a 'handmaiden', and is assigned to a 'Commander' to bear his children; she has no value beyond childbirth, with the impending threat of being shipped to the 'colonies' (the typical radioactive dump every dystopian book is incomplete without) on her failure to get pregnant looming over her. She is referred to as Offred (Of-Fred) which serves to highlight how replaceable she is; individual identity being non-existent. It was precisely at this point when I realized the Handmaid's names (Offreds, Ofwarrens, and Ofglens) that a slight jolt went through me. Hence far I'd read with a cool detachment, surely it was too exaggerated and distorted a version of society to take personally?

But, wasn't Offred eerily similar to the modern Mrs—to take up your husband's name and leave your own identity behind? We normalize it, of course, wrap it up under the pretty farce of being united as one family, but isn't the core concept of controlling women the same, and wouldn't a man be outraged at the prospect of taking up his wife's name?

This was the first time Atwood did what she does countless times through the pages of *The Handmaids Tale*, drawing light to the absolutely shocking gap between both genders even today without openly talking about them. She utilizes a powerful tool, *empathy*. While the average person might joke about 'triggered feminists', or how it's an outdated movement, one can't help but feel for Offred; to experience her life with a sickening horror further augmented by the realization that it's not as different from present-day as you thought it was. It's easy to say that women today have equal rights, and dismiss the feminist movement as obsolete, but it is jarring to see women treated not very differently than they were mere decades ago. Not treated in a literal sense, but it was a mutual understanding among both men and women that women are naught but a means for reproduction. Being infertile was the lowest of disgraces, and having borne and raised multiple children garnered respect. Even in Nazi Germany, women were awarded 'crosses' for childbearing: gold for seven, silver for six, and bronze for five. A trophy. Something to wrap in shiny ribbons. Offred strips those ribbons away, offering a candid glimpse into how women were perceived, and how brutally inhumane that perception was.

Offred gives us flashes of both her life (more like predicament) and society; every movement is preordained, every hour of her life accounted for, quoting the book *"A rat in a maze is free to go anywhere, as long as it stays inside the maze."* But lack of free will is characteristic of many a dystopia; a political system that is objectively agreed to be 'bad', a system that readers are aghast by, and isn't really explored in a way that makes a person go "oh wow what if we lived like this? Who's to say this wouldn't be a better system" or something along those lines. It's always "thank goodness we don't live like that", Atwood, however, gives us and rationalizes (somewhat similar to Orwell in 1984) the view of the ominous 'They' rather than presenting 'Them' as cold, mean, and inhumane. Examining a few excerpts from the book to elucidate my point:

"There is more than one kind of freedom," said Aunt Lydia. "Freedom to and freedom from. In the days of anarchy, it was freedom to. Now you are being given freedom from. Don't underrate it.

"We've given them more than we've taken away, said the Commander. Think of the trouble they had before. Don't you remember the singles bars, the indignity of high school blind dates? The meat market. Don't you remember the terrible gap between the ones who could get a man easily and the ones who couldn't? Some of them were desperate, they starved themselves thin or pumped their breasts full of silicone. Think of the human misery."

But you also find spurts like this from Offred; which brings back a much saner perspective, saner in the sense that you can recognize the cruel apathy of the society Atwood paints:

"All alone by the telephone. Except I can't use the telephone. And if I could, who would I call?

Oh God. It's no joke. Oh God oh God. How can I keep on living?"

While going back to a sexist society is unlikely, the demarcation of gender roles—typically with women being considered inferior to men—is prevalent in many parts of the world, particularly third world countries, with girls being married off as young as fifteen, their sole purpose being to run a household and bear sons. Offred reminds us to not trivialize their suffering, to not accept their position as an unfortunate byproduct of a flawed generational mindset.

Perhaps Atwood wrote a cautionary tale to warn us against extreme right-wing religious ideology and the kind of society it might produce, or perhaps a social critique on the current state of affairs through showing a brusquer version of the same, either way, it hit its mark.

A brilliant and novel addition to both feminist literature and science fiction, The Handmaid's Tale takes the objectification of women to an entirely different level, making us question today's society, and take women's rights movements more seriously.

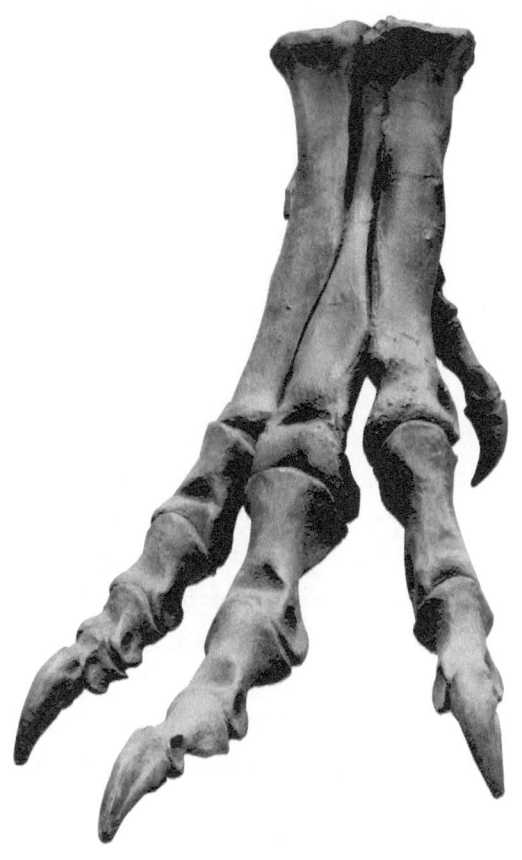

CHAPTER 7

JURASSIC PARK

Michael Crichton, 1990

An award-winning movie by Steven Spielberg, and a book by Michael Crichton which we don't give due credit to, *Jurassic Park* became a cultural icon soon after its release, a thriller crafted on the underlying premise of a theme park with dinosaurs. Doesn't sound too impressive? Understandable. There have been far more audacious plots in the ever-growing realm that is science fiction, but for a book written in 1990 (before, yet not too long before the first animal was cloned, Dolly the sheep in 1996), Jurassic Park came to be the mascot of genetic engineering and drew popular attention towards biotech, or biotechnology, as a subject.

Let me rewind(/digress) slightly.

On one of my routine midnight rants with a friend, we were discussing robot pride. Pride parades for people falling in love with robots, or robots falling in love with each other; that perhaps that would be liberal twitter in another four decades. We didn't talk about robots per se, that's overdone and dreary, yet our conversation inadvertently revolved around robots. In a similar fashion, Crichton doesn't talk exclusively about making dinosaurs, or present it as the holy grail; he instead chooses to focus on *why* they were made, and *what* happened after without skipping on *how* they were made.

The plot goes thus: John Hammond, an ambitious and wealthy man sets out to create 'something real', namely Jurassic Park, a theme park with dinosaurs. Dinosaur DNA (Deoxyribonucleic acid) is obtained from mosquitoes found frozen in amber, which naturally sucked dinosaur blood centuries back. Hammond buys up amber deposits in huge deposits, till he finds the aforementioned mosquitoes. Sparing no expense (his motto, we later find out) Hammond assembles a top-notch team, backed by affluent investors, to put together the boldest theme park in history. Enter Alan Grant (a paleontologist), Ellie Sattler (a paleobotanist), Ian Malcolm (a mathematician), and a horde of other characters I am not going to name, who come to visit the park in its final stages of construction on Hammond's request. Enter Dennis Nedry, a programmer hired by Hammond who is being paid heavily by a third party to steal dinosaur embryos from the park. Nedry shuts down the park, including communication lines as well as the electric fences, and attempts (has an unfortunate encounter with a dilophosaurus along the way) to flee the island with the embryos.

Meanwhile, the rest of our characters are stranded on an island with most of the communication and safety systems down and dinosaurs on the loose.

While it naturally makes for an engrossing narrative, with a Tyrannosaurus Rex and Velociraptors, with people casually being eaten, and a literal 'fight-for-survival' vibe foreshadowing the whole book, it is surprising that many people wholeheartedly believed that resurrecting (or recreating) dinosaurs was inevitable. After all, the book made it seem so easy— get hold of some dinosaur DNA, fill in the gaps where needed, do some fancy sciencey stuff, and there you have it!

It is easy to get swayed by what's impossible and what's not with dinner conversations turning to 'technology these days' followed by an incredulous shake of the head, with the line between science and fiction getting blurrier every day, with pessimism being taken for conservatism or backward thinking. Everything is possible with science, technology and software seem to be the common refrain. But at a much more practical level, it really isn't that hard to distinguish between what may one day be possible, and what has a clear 'no' factor that we can't get around.

Biotech is making some truly remarkable advancements because gene splicing and genetic engineering are fields growing at an alarming rate, but reviving history? Jurassic Park is scary not just because of the constant adrenaline coursing through our characters, but because it seems like an unusual yet not entirely implausible scenario. It is ordinary people as amazed by dinosaurs and what Hammond has created as we are, caught in a nightmarish situation. It is scary because people at some level believe that Jurassic Park might someday be a reality—hopefully with better security protocols.

Now coming to the big question; is Jurassic Park happening anytime soon? The short answer is probably not, and the long answer is also probably not. Crichton said it himself:

"The problem was that all known dinosaurs were fossils, and the fossilization [had] destroyed most DNA.... So cloning was therefore impossible. There was nothing to clone from. All the modern genetic technology was useless. It was like having a Xerox copier but nothing to copy with it."

Crichton got around this obstacle of course, in his mosquito-embedded-in-amber scenario, but DNA does degrade in about 10,000 years, and extracting it from something millions of years old, and finding such DNA intact after years of geological upheaval isn't realistic, or possible as of today.

However, Jurassic Park isn't solely about dinosaurs and genetic modification. It's about the consequences of capitalistic greed, and the consequences of the intrinsic human desire to interfere with natural hierarchy. As a species, humans have risen to the top of the food chain through innovation and persist in their efforts to break more barriers and restructure biological hierarchies to their advantage. Hammond is a classic example of this human arrogance, he pays no regard to the ethicality of what he is doing, it is all about the money and grandeur for him. Malcolm warns Hammond that the park is inherently unsustainable, and man can't tamper with a natural selection without repercussions, yet no one else shares his concerns. The system does eventually collapse due to factors that Hammond couldn't have foreseen, or controlled—human greed, and nature. Quoting Crichton,*Life finds a way.*

This recognition of the powers of genetic engineering and simultaneous skepticism accompanying it are startlingly relevant in a modern context, especially with respect to the recent breakthroughs in biotech.

CRISPR (Clustered Regularly Interspaced Short Palindromic Repeat) is a family of DNA sequences found in the genomes of prokaryotic organisms such as bacteria and archaea. These sequences are derived from DNA fragments of bacteriophages that had previously infected the prokaryote. They are used to detect and destroy DNA from similar bacteriophages during subsequent infections, as per Wikipedia. CRISPR sequences were first discovered in 1987 (three years before *Jurassic Park* was released), and this system was first used in 2008. CRISPR won a Nobel Prize in 2020, due to the breakthrough research within this field by Jennifer Doudna and Emmanuelle Charpentier. CRISPR Cas-9 essentially allows us to edit parts of a genome by altering certain segments of a DNA segment. This has wide-ranging, and potentially serious, implications. Genetic mutation can allow us to 'modify' various organisms—including humans—to incorporate certain desirable traits. Although trials are strictly regulated, and the ethicality and legality of future human trials is under debate, CRISPR Cas-9 has opened up an avenue of possibilities that were historically viewed as pure science fiction.

While most of the public is fascinated, though somewhat trepidant, by the opportunities CRISPR-Cas9 presents, the eerie similarity of this scientific breakthrough to a pop-culture phenomenon based on the same grounds—genetic engineering—which debuted shortly before it and elicited the same reaction from both its characters and the public is not to be ignored.

Although we obviously can't base our opinions on a piece of fiction, Crichton's cautionary words faintly echo every YouTube video and article on genetic engineering. Life finds a way. Do not tamper with it. The last twenty years have seen rapid advancements in genetic, biomedical electronics, robotics and AI and it is widely expected that capabilities to alter various life forms will get quite mature in the next few decades. The implications of artificially enhanced humans, animals or plants are complex, potentially dangerous and will set humans on a new uncharted path.

In a speech to Congress in September 2006, Crichton pleaded the various statesmen present, to not meddle with the nuances of genetics. He further claimed that "we have plenty of evidence that today, gene patents are bad practice, harmful and dangerous. End that practice now." Continuing to assert his argument, "It's all a mess. And it's a dangerous mess....We need added legislation to clarify a legal conception of human tissues and how they are used. Federal rules already exist, but the courts are ignoring those rules, and are confused because they are trying to reason based on prior property law....We need clear laws." Crichton also claims that an absence of thorough research in the field of genetic engineering will lead to heinous mistakes and can even lead the world to the brink of an apocalypse.

Crichton's take on the subject may be an exaggeration, but caution, humility, and respect for natural selection are essential to the next steps in genetics; *Jurassic Park* beautifully illustrates how control over nature is an illusion. No matter how scientifically sound a procedure may seem, and no matter how much money and technology we pour into a project, reckless advent into and interferences in natural biological structures can have dire, unprecedented outcomes.

CHAPTER 8

SNOW CRASH

Neal Stephenson, 1992

Written in 1992 and read in 2020, Neal Stephenson's *Snow Crash* offers a wonderful introduction to Cyberpunk—a subgenre of science fiction that envisions high technological achievement but a broken-down, gritty society. Hence the 'cyber' and the 'punk' – also often summarised as 'high-tech, low-life'. These don't envisage a shining future of intergalactic civilizations—such as *The Hitchhiker's Guide to the Galaxy*, *Star Wars*, or *Star Trek*—nor are they as pessimistic as Orwell's *1984* or Huxley's *A Brave New World*. Set in the near-distant future, cyberpunk plays on conflicts between hackers, AI surveillance, futuristic technology and mega-corporations. Cyberpunk is particularly interesting and relevant today, as we tilt towards a fragmented society whose primary objective is peace among nations and corporations, not a unified world state. We face the early challenges of interconnected societies, individual privacy, and the role of governments versus Big Tech, and cyberpunk is invaluable to chart out possible routes these conflicts can take. Additionally, the emergence of freelancers and groups who feel that Silicon Valley is elitist and seek to redefine the tech circuit augments this uncertainty regarding the future of technology and its role in society.

Snow Crash is set in future Los Angeles amidst an anarcho-capitalist America driven by consumerism where corporations run everything from the police to highways. The USA has fragmented into multiple franchises run city-states each with its own laws – these need visas to travel to and from. The federal government retains its bureaucracy in theory, but has lost a significant proportion of its power. All in all, the capitalist dream that Big Tech pushes for today. But Stephenson reveals the impact this ideal can have in society, further cultivating the individualist mindset already prevalent in Silicon Valley among the people, and creating a society with no real sense of or allegiance to 'society'.

Throw in the 'Metaverse' – a cross between virtual reality and the internet, forming a virtual environment where people can log in and hang out as Avatars in virtual cities complete with roads, cars, public transport and exclusive clubs.

Coming to the characters, Hiro Protagonist—yes, the hero and the protagonist of this book—delivers pizzas – a very serious business handled by the Mafia Franchise. In addition to being an excellent sword fighter, Hiro is a freelance hacker and one of the original designers of Metaverse, carrying considerable clout there. The book starts with Hiro at risk of missing a delivery deadline but being saved by Y.T, a stereotypical 'punk' teenager aboard her skateboard. The antagonist, Bob Rife, is a telecommunication & computer tycoon seeking world domination through a virus called Snow Crash.

The Metaverse is littered with dogfights and gore—after all, people can't really die there—till the arrival of a new virus – Snow Crash.

Stephenson comes up with a concept of neurolinguistic hacking, where language is used as programming for the brain. People become zombies by watching some bitmap on-screen or listening to some ancient Sumerian words. It requires suspending some belief, but Stephenson works hard to make it plausible by blending in religion, linguistics, anthropology, and mass communication as ways to influence minds. It's not that alien either, just an exaggerated version of what social networks are capable of doing today—think the Myanmar Military, the Rohingya genocide, and Facebook. And that is only one of the publicized incidents, multiple more happen on a minuscule scale.

A conversation between Hiro and his ex-girlfriend who recruits him to fight Bob Rife goes as follows:

"Wait a minute, Juanita. Make up your mind. This Snow Crash thing—is it a virus, a drug, or a religion?"

Juanita shrugs. "What's the difference?"

Another key character is Raven, a giant Aleut who rides a motorcycle and uses glass knives to rip people to pieces. He has reasons to hate America, including the Nagasaki bombing that killed his father.

Snow Crash is amazing in terms of the breadth of concept coverage, integrating computer science, history, anthropology, politics, philosophy, and more. The writing switches from fast-paced action scenes to explanations.

The book is written in the present tense, leading to a quirky narrative. Some passages are gems that would be cult-classics if *Snow Crash* were ever developed as a movie:

"once we've brain-drained all our technology into other countries....there's only four things we do better than anyone else: music, movies, microcode (software), high-speed pizza delivery"

"The world is full of power and energy and a person can go far by just skimming off a tiny bit of it."

"They are all done up in their wildest and fanciest avatars, hoping that Da5id—The Black Sun's owner and hacker-in-chief—will invite them inside. They flicker and merge together into a hysterical wall. Stunningly beautiful women, computer-airbrushed and retouched at seventy-two frames a second, like Playboy pinups turned three-dimensional, these are would-be actresses hoping to be discovered."

"Until a man is twenty-five, he still thinks, every so often, that under the right circumstances he could be the baddest motherf*cker in the world. If I moved to a martial arts monastery in China and studied real hard for ten years. If my family was wiped out by Colombian drug dealers and I swore myself to revenge. If I got a fatal disease, had one year to live, and devoted it to wiping out street crime. If I just dropped out and devoted my life to being bad.

Hiro used to feel this way, too, but then he ran into Raven. In a way, this was liberating. He no longer has to worry about being the baddest motherf*cker in the world. The position is taken."

But I digress.

Coming to the cons, despite the potential the plot had and the witty writing style, Stephenson could have done a much better job with plot development, dragging on towards the end. The narrative has a certain light-heartedness, which was perhaps meant to remind the reader not to take it too seriously. However, the flippancy and satire clash with serious, complex explanations of the Sumerian namshubs, linguistics, refugees on the raft and the imminent infocalypse. Throw in casual racist references while describing the burbclaves, the refugees, and the raft, and you're left with a slightly confused reader.

Somehow, over the pages, the story, the characters & the outcome lose importance and reader engagement dips as the plot gets flaky & unreal. Overload & complexity of concepts takes away the soul from the book, making the last 100 pages feel dreary and a chore to be finished. Considering the book has only a few characters, it was surprising that it could not make me particularly care for most of them. Maybe that's how a book about cyberpunks & a laissez-faire society in a dystopian future should be – but it got a bit impersonal for my taste.

Having said that, *Snow Crash* is a seminal book on cyberpunk and established Neal Stephenson as a futurist in Silicon Valley. While the idea of the Internet and Avatars was not new when the book was written, *Snow Crash* made it larger than life and inspired a bunch of developments over the next few years. From the video game Quake to applications like Second Life & Myspace and of course, the movie Matrix, all to some extent, were inspired by Snow Crash. Even Google Earth bears a resemblance to the "Earth" software developed by the Central Intelligence Corporation in *Snow Crash*. Other publications, such as Ready Player One too employed a similar concept of a virtual reality-internet hybrid after *Snow Crash*.

Additionally, humans being 'hacked', while brought out differently in the book, bears an uncanny resemblance to tech giants such as Facebook pushing us into echo chambers, and reconfiguring our perspective through targeted content. The core concept of getting inside human brains is in effect today, and *Snow Crash* shows us what it might have looked like had technology gone down a different path.

The other interesting take is on currency, where people transact in untaxable electronic currency, which leads to a decline in government tax revenues and its power. This leads to the devaluation of government printed currency so that quadrillion dollar bills are being used as small change. Sounds like cryptocurrency and how governments around the world are trying to figure out a way to regulate and deal with it.

It is interesting to note that Stephenson wrote the book in 1992 — email had then just started and the Internet boom was still to arrive. Stephenson's genius lay in being able to imagine a dystopian hyperconnected society, infatuated with virtual avatars, where global tech corporations wield more power than governments. At its core, Snow Crash is thus a good—albeit slightly pessimistic—reflection of the tech evolution and how it could play out. The fun writing style and the cool gadgets not only compensate for the lackluster plot structure, but also subvert the underlying theme of a dystopian society torn apart by the power play, with technology it still can't quite control.

By and by as I compare the metaverse with the way we as a society are progressing, the same question echoes—are we biting off more than we can chew?

CHAPTER 9

THE MARTIAN

Andy Weir, 2011

"The one starring Matt Damon which was nominated for an Oscar, right?" No, the one written by Andy Weir. Quoting the author, it turned out to be a 'thriller which could double as a science textbook'.

The story revolves around Mark Watney, our protagonist, and traces the efforts it takes to get him home. Watney is a botanist-astronaut who is blown away by a storm while collecting samples from the Mars surface. The commander of the mission assumes that he wouldn't have survived the falling debris, and his crew heads back to earth without him. Thus, Watney ends up stranded on Mars, with a limited number of rations. He has a clear, but daunting task ahead of him: making contact with the base back on earth, and, of course, trying to survive as long as he can before being rescued. Being a botanist helps, as Watney grows food using his own biowaste, and sets up a potato farm inside the vessel. He deploys basic science to make water by combining hydrogen and oxygen. Watney does eventually manages to make contact with the earth, and his crew goes back on a mission to rescue him. In all, he spends over five hundred days alone on Mars before getting picked up for his ride back home to earth.

Watney faces challenges that seem insurmountable —lack of life support including water and food, and the omnipresent risk of equipment failure —yet finds ingenious ways of pulling through. Weir gives us a witty hero, a novel plot, and a setting that doesn't fail to excite readers. But what's truly remarkable about *The Martian* is the degree of precision, as well as the attention to detail. The plot isn't a cluster of 'sciencey' terms or have 'quantum physics' as an excuse for major plot holes, but it is almost as though Weir intended to write something along the lines of 'Mars: How to get there and how to survive', and threw in characters for the heck of it. Through multiple narratives and perspectives, the author gives us a steady picture of how the hypothetical Mars Missions, or the 'Ares' missions, work without compromising on feasibility. From calculating the trajectories and days required to traverse from one point to specifying how a tarp is made, Weir lets nothing slide unresolved. But even with all the figures, it's a thoroughly engaging book littered with humor, and sufficient diversity in character point-of-views.

Published in 2011 —at the cusp of an era of privatization of space as SpaceX and Blue Origin gradually started gaining momentum —*The Martian* certainly contributed to the hype of Mars Missions. It drew in public attention by touching upon this field of space exploration without involving time travel, aliens, or other elements incorporated for the 'thrill of it' to make for an exciting narrative. Rather, Weir made the mission seem entirely probable in a not-too-distant timeframe and gave us a dream to pursue, a catalyst of sorts for the future Mars missions.

In practical terms, *The Martian* is a slight digression from typical science fiction that usually focuses on ideas and ignores functionality (take *Frankenstein*, Shelley never told us *how* he made his monster). Weir's idea is relatively easy in theory —Mars mission —and he instead chooses to elaborate on the functioning. It is, however, a revolutionary book, as it outlines a full-fledged manned mission to Mars —something that SpaceX by Elon Musk and Blue Origin by Jeff Bezos has been looking to do towards the end of the 2020s. The Martian, set in 2035 offers the perfect guidebook (or perhaps a better way of think about it would be what-not-to-do-while-going-to-Mars) to do so, as it is the first time that fiction in this field has been written on this sort of a scale, and doesn't employ technologies whose acquaintance we haven't yet had the privilege of making.

Out of the technologies Weir does employ, most of them are already being implemented at NASA (take the example of the 'hab', Watney's 'habitation module', similar to NASA's 'HERA' or Human Exploration Research Analogue), and there isn't too much 'innovation' per se. But where Weir lacks innovation in terms of technology, he compensates in plot structure. He takes these disjoint pieces of research on survival on Mars to construct his jigsaw, giving clarity on how they might fit together and a broad outline on the challenges that may arise in putting together a Mars mission. He doesn't piece together the entire puzzle, but the parts he pieces do work. They are theoretically sound. And that's not something a lot of —or any book in this genre on this subject—can claim.

Don't get me wrong, The Martian isn't a piece of ground-breaking research that pre-empts how the actual (and inevitable) Mars missions will play out, although it does firmly reinforce the inevitability of the missions themselves. There are plenty of elements that are exaggerated and exist purely to add to the 'thrill' aspect, for example, dust storms on Mars aren't potent enough to knock over a spacecraft or wreck damage to the degree they did in the book. But even as a thriller, while Watney is likable he comes off as uni-faceted.

Psychologically, Weir fails to incorporate the range of emotions, and trauma Watney must go through; being the only human stranded on an alien planet with a very low chance of survival is bound to be traumatizing. All we're exposed to is whatever dire situation crops us, an ingenious solution, and humorous remarks. And repeat. But I digress.

The Martian is a tribute to science, mathematics, innovation, and human perseverance. It showcases the scope of fields and skillsets science and technology includes, and touches upon their interdisciplinary nature. Space travel and space survival requires the smooth integration of multiple fields—from botany to computer science to astronomy to chemistry, as Watney demonstrates. By weaving together a convincing story with interesting characters, Weir successfully brought space exploration back into mainstream media, and reinforces that Mars missions are the next step towards pushing the boundaries of science, technology, and innovation.

Coming to actual manned Mars missions, the 'science' is catching up to the 'fiction'. They're no longer a distant dream (distant, yes—roughly 345.6 million kilometers distant), but an eventuality. As a species, we're technologically capable of reaching Mars, and colonizing Mars is the first step towards an intergalactic civilization, which is supposedly necessary for our long-term survival. But that's way ahead. As of December 2020, Elon Musk is 'highly confident' that SpaceX will ensure humans set foot on Mars by 2026. SpaceX has grown exponentially and made remarkable progress over the past decade, achieving the lowest launch costs in the history of space-flight. These achievements can be accredited to its novel method of carrying out most of the manufacturing within a singular facility, with low dependence on external contractors, and by using recycled rockets.

Overall, the Martian is geeky, technical, and optimistic. But it has fueled the imagination of millions and is part of the prelude to actual Mars Missions.

CHAPTER 10

READY PLAYER ONE

Ernest Cline, 2011

Science fiction upholds its tradition of looking into the future, and coming back with various questions for humanity, in Ernest Cline's *Ready Player One*. Some cliched, yet relevant. Where will technology take us? To what extent will humans integrate themselves with their virtual personas? Articles on 'social media addictions' are already an eyesore, it is undeniable that virtual identities are no longer charades to be discarded at leisure, they're gradually becoming as important (if not more important) as our real-life ones. Identity and an intrinsic sense of self-worth revolve around how we present ourselves on social media, and society's reaction to it. It is convenient to reinvent yourself and project a version of who you want to be rather than who you are, however, the strong real-world dissociation accompanying this quest for a picture-perfect Instagram is an effect of the same. But is this effect truly a negative one?

Ready Player One explores these questions through OASIS (Ontologically Anthropocentric Sensory Immersive Simulation), a virtual reality cross social media hybrid. Set in 2045, Cline envisions a future where humanity is overwhelmed with many of the issues that threaten today's society, such as climate change, a global energy crisis, and overpopulation. Apathetic citizens become acclimatized to their laissez-faire society, and the poor conditions they live in.

Rather than improving their circumstances, people immerse themselves in the OASIS, building a new life for themselves in an unregulated world where anything (literally) is possible. Cline thus cleverly touches upon one of the primary motifs driving humanity: escapism.

Escapism isn't a new concept. Finding ways to drown out the monotony of our lives, to 'escape reality', is a skill that humans have continually honed and perfected. Books turned to movies, movies to video games, and video games to virtual reality, each decade and the technologies accompanying it promised an even more engaging, immersive way to shut out the real world—for a short period. A very short period. But as Social Media crept upon us, starting with MySpace and Friendster and truly taking off with Facebook, we were no longer the solitary characters in our makeshift worlds, and we could stay there as long as we liked. The foregoing of reality for a fabrication that does have real-world implications doesn't sound particularly appealing at the surface, but inexplicably draws people in by capitalizing on this inherent human instinct to escape. As virtual reality and gaming gain momentum in parallel to this spike in societal connectivity, a synthesis of both seems inevitable. Cline demonstrates what this may look like through the OASIS. The OASIS is essentially a virtual simulation that anyone can access using a 'hapsuit' and a 'visor'—similar to a VR headset, but considerably more sophisticated—and interact in. The internet, but on steroids.

Our protagonist, Wade Watts, lives in the OASIS. An overweight, acne-ridden, broke orphan, he can escape his circumstances through the OASIS, taking on a new avatar ('parzival'), and socializing with strangers. Quoting Wade,

"Online, I didn't have a problem talking to people or making friends. But in the real world, interacting with other people—especially kids my own age—made me a nervous wreck. I never knew how to act or what to say, and when I did work up the courage to speak, I always seemed to say the wrong thing."

And a little further, we have,

"On my first day at OPS #1873, I thought I'd died and gone to heaven. Now, instead of running a gauntlet of bullies and drug addicts on my walk to school each morning, I went straight to my hideout and stayed there all day. Best of all, in the OASIS, no one could tell that I was fat, that I had acne, or that I wore the same shabby clothes every week. Bullies couldn't pelt me with spitballs, give me atomic wedgies, or pummel me by the bike rack after school. No one could even touch me. In here, I was safe."

It's natural that Wade would want to escape his woes through the OASIS, yet ironically augments them by neglecting his fitness and well-being in the real world. Is this necessarily bad though? While the reflexive answer might be 'yes', this scenario isn't as alien as it seems. After all, don't we meticulously arrange our calendars to showcase a vibrant social life on social media? Rather than capturing the good and sharing it with the world, we do good for the sake of sharing it with the world. Cline doesn't give us an immediate verdict on this virtual-world obsession being 'good' or 'bad', he just shows us a mirror.

Coming to the plot: the creator of the OASIS is James Halliday, a game designer obsessed with the 80s, and when Halliday dies in 2040, he leaves his fortune—worth trillions—to anyone who can find his Easter egg, hidden deep in this parallel universe.

Thus Halliday, even after his death, remains alive in the hearts and minds of all the gunters (nickname for the egg hunters), who start to pour over every 80's pop culture reference, no matter how big or small. Halliday scholars exist, keeping his spirit alive by reviving whatever Halliday found interesting. Wade too becomes one of the many fans and idolizers of James Halliday, scouring and devouring everything about the decade of 1980, both for research and as a way to escape the futile existence in a near-apocalyptic rendering of planet earth, which is almost unfit for living. And thus Cline unfolds a theme echoing throughout his novel—nostalgia.

It's interesting to see that one of the central themes of Ready Player One is nostalgia, particularly while living in a generation that is eager to reach the future, and learn from, but not cling to, the past. Nostalgia in *Ready Player*

One, is not for a past life in the novel's contemporary world, but for a world long lost. As gunters devour pop anthologies pertaining to the 1980s, they not only revive the spirit of that decade but also visualize the world as once it was: a far better and cleaner place. This longing for the past motivates James Halliday to design his hometown Middleton as a planet, where he hides the first gate of the quest. And as Wade points out, the painstaking attention to detail that Halliday puts into while creating that virtual world that is the simulation of Haliday's teenage years, is mind-blowing. But it is also indicative of the way Halliday tries to preserve those cherished memories of his favorite games, bowling alleys, and a town that is long lost.

Perhaps Cline seeks to highlight how good we have it today, and how dangerously we toe the precipice of losing it as we relentlessly debate pressing global issues such as climate change instead of solving them. Or maybe it just makes for a more gripping narrative. While Wade and his team compete against innumerous other players, the true antagonist of the novel, however, is capitalism. IOI (Innovative Online Industries)—the world's largest internet service provider and telecommunications company—is also fixated on Halliday's legacy. IOI hopes to take over the OASIS, and monetize on it, throw in a monthly user fee, advertisements, and access to customer data, by exploiting Halliday's will.

As Cline spins a riveting plot peppered with pop-culture references, he also alludes to the dangers that profit-driven ventures pose in an unregulated environment. This warning isn't baseless; capitalism and ethicality in tech have been touchy subjects. The promise of 'connectivity' and 'broadening our worldview' lure us to social media, but social media paradoxically pushes us deeper into our echo chambers and perpetuates our confirmation bias, allowing Big Tech to profit off the highly polarized society we are now becoming. The Cambridge Analytica Scandal, for example, reveals how valuable a currency customer data is, and the implications it can have by allowing tech tycoons and politicians to use a person's psychology against them. There is still a need for greater awareness, stricter data privacy laws, and an emphasis on ethical design within tech.

Wade's team (nicknamed the 'High Five) include Aech, a black, gay teenager whose mother disowned her, Art3mis, whose body dysmorphia stems from a large birthmark on her, Daito, a Hikikomori, and Shoto, Daito's surrogate sibling, and only companion. They inevitably flock to the OASIS to avoid the social stigma they face in real life, and as Cline presents a virtual world where everything can be perfect, he simultaneously draws attention to eliminating the need for such a world by making the real one a more accepting place. *Ready Player One*thus, despite the fast-paced plot set in a vibrant world, paints a bleak picture of what might become of the real world if current issues such as climate change and social stigma persist. Rather, if we allow them to persist.

A substantial part of*Ready Player One*, is a celebration of video games and the importance of pop culture in an individual's life.

OASIS becomes a place where anyone can be anything, no matter who they are, and they can learn everything about anything, with infinite knowledge available at their fingertips. In order to win the game, Wade must love playing the game, whether it is a high-definition simulation of reality or a one-dimensional game of the 1950s. Wade is the perfect successor to Halliday, due to their shared love of video games, and pop culture. All in all, Ready Player One is an important addition to science fiction, highlighting the role of capitalism in a futuristic simulation of a three-dimensional version of the internet, as well as the repercussions of giving virtual identities more importance than our real ones. But it also brings out softer themes—of staying attached to what we love and not giving it up in our quest for profiting off technological advancement.

Concomitantly, Cline's primary focus isn't to draw attention to the abysmal state of the real world (in his world). It is to give us a story, with a very strong, and relevant moral:

I'm not crazy about reality, but it's the only place to get a decent meal.

While technological advancement is a good thing, real human feeling and contact trumps everything and every digital simulation, and therefore the last message of the book is where Wade realizes he doesn't want to log back into OASIS because he has Art3mis with him in person, the love of his life. This points to what Halliday has intended people to understand: you might play games and it's online where you are safe from the outside world, but humans need humans and the earlier we realize this, the better.

Not getting disconnected from reality in our quest to get more 'connected' isn't anything new, but that doesn't make the thought any less profound. It is a sentiment not as widely shared since COVID-19, reprimands about "too much screen time" are met by retorts of "I need it for school", as we have come to accept our over-dependence on technology. There's boundless optimism for where innovation and technological advancements will take us, but also a plea to not neglect the real world. Cheesy and predictable, Cline gives us a cliché towards the end—heroes emerge victorious, 'greedy-evil-capitalist' goes to jail, guy gets girl, the world is less OASIS obsessed—but it is a cliché done well. And as we're left with thoughts of how ethical design, capitalism, virtual personas, and escapism are crucial keywords to shaping our future, there's a stronger thought foreshadowing them—hope.

About The Author

Damini Rana is currently a senior in Delhi Public School RK Puram, graduating in 2022. A science student fond of literature, she is interested in observing and understanding the social shaping of the world through technology. When not immersed in science fiction, she can usually be found eating pancakes, obsessing over clean typography, listening to *The Strokes*, dancing, or reading fanfiction.